Integrated Converters

Series Editors
G. Lancaster J.K. Sykulski E.W. Williams

1. Introduction to fields and circuits
   GORDON LANCASTER

2. The CD-ROM and optical recording systems
   E.W. WILLIAMS

3. Engineering electromagnetism: physical processes and computation
   P. HAMMOND and J.K. SYKULSKI

4. Integrated circuit engineering
   L.J. HERBST

5. Electrical circuits and systems: an introduction for engineers and physical scientists
   A.M. HOWATSON

6. Applied numerical modelling for engineers
   D. DE COGAN and A. DE COGAN

7. Electromagnetism for engineers: an introductory course (Fourth edition)
   P. HAMMOND

8. Instrumentation for engineers and scientists
   JOHN TURNER and MARTYN HILL

9. Introduction to error control codes
   SALVATORE GRAVANO

10 Materials science for electrical and electronic engineers
   IAN P. JONES

11. Integrated converters
    PAUL G.A. JESPERS

Integrated Converters

D to A and A to D ARCHITECTURES, ANALYSIS AND SIMULATION

Paul G.A. Jespers

Université Catholique de Louvain, Belgium

OXFORD
UNIVERSITY PRESS

Great Clarendon Street, Oxford OX2 6DP

Oxford University Press is a department of the University of Oxford.
It furthers the University's objective of excellence in research, scholarship,
and education by publishing worldwide in

Oxford New York

Athens Auckland Bangkok Bogotá Buenos Aires Calcutta
Cape Town Chennai Dar es Salaam Delhi Florence Hong Kong Istanbul
Karachi Kuala Lumpur Madrid Melbourne Mexico City Mumbai
Nairobi Paris São Paulo Singapore Taipei Tokyo Toronto Warsaw

with associated companies in Berlin Ibadan

Oxford is a registered trade mark of Oxford University Press
in the UK and in certain other countries

Published in the United States
by Oxford University Press Inc., New York

© Paul G.A. Jespers 2001

The moral rights of the author have been asserted
Database right Oxford University Press (maker)

First published 2001

All rights reserved. No part of this publication may be reproduced,
stored in a retrieval system, or transmitted, in any form or by any means,
without the prior permission in writing of Oxford University Press,
or as expressly permitted by law, or under terms agreed with the appropriate
reprographics rights organization. Enquiries concerning reproduction
outside the scope of the above should be sent to the Rights Department,
Oxford University Press, at the address above

You must not circulate this book in any other binding or cover
and you must impose this same condition on any acquirer

A catalogue record for this book is available from the British Library

Library of Congress Cataloging in Publication Data
(Data applied for)

ISBN 0 19 856446 5

Typeset by EXPO Holdings, Malaysia
Printed in Great Britain
on acid-free paper by Biddles Ltd., Guildford, Surrey

To Denise

Preface

Digital technologies aim towards ever smaller, faster, less consuming transistors to lead to increasingly complex integrated digital systems. Consequently, many of the functions obtained through the use of analog integrated circuits are now entrusted to digital circuits. This has reaped substantial benefits regarding accuracy with no sensible penalty as far as area and power consumption are concerned. The input data are often analog however so that high performance imbedded D to A and A to D converters are required unless already integrated within the sensors linking the system to the outside world.

Converters are essential parts that enable communication between the external analog world and the digital silicon chip. They should not compromise precision even though the hardware in which they are implemented relies on semi-conductor devices known for their poor accuracy. Therefore, converters capitalize design expertise accumulated during the last 20 years to circumvent the limitations and impairments inherent to integrated circuits.

The objective of this book is not only to review the principal converter architectures, but also to bring to forward many of the innovative solutions suggested throughout time to reach high performance. This is why some circuits, nowadays obsolete, but essential stepping stones in the process of development, are still reviewed. Their merit lies in their ability to illustrate the basic principles upon which progress became possible. To this end, in addition to detailed presentations of the various types of converters, this book also exemplifies the evolution designers went through in order to cope with the inherent integration limitations. Converters offer a remarkable opportunity in this respect. They make the most of the hardware and software techniques available to enhance the performances of integrated circuits.

The book is divided in eight chapters. Definitions and evaluation techniques are dealt with in Chapter 1; parallel D to A converters in Chapters 2 and 3; serial A to D converters in Chapters 4, 5 and 6; stochastic A to D and D to A converters in Chapter 7; and flash, multi-step, pipelined and folding A to D converters in Chapter 8.

Scaled D to A converters use binary or linearly weighted current references, voltages or charges, taking advantage of quasi-randomly accessed unit-elements to average out mismatches. Their accuracy is hardly better than 10-bits. Other approaches are essential for higher resolution. The segment converters dealt with in Chapter 3 offer a good trade-off, although their linearity calls for very accurate segment references. High accuracy converters may also be implemented by taking advantage of dynamic current matching techniques such as those considered in the first part of Chapter 3. Other D to A converters are reviewed in the remaining chapters, although the emphasis is put on A to D converters. These are more complex than D to A converters. True parallelism requires large area and consumes a lot of power. A number of distinct approaches sacrificing speed for smaller silicon area and power consumption are reviewed in Chapter 8. These combine series and parallel techniques. Series architectures derive from the single-bit serial converters considered in Chapters 4 and 5. The converters currently designated as subranging, recycling, pipelined and folding converters, fill the gap between fully parallel and serial devices, capitalizing many of the techniques described in Chapters 2 to 5. Most offer remarkable speed performance, reaching tens of MS/s sampling rates. Few of these converters achieve an accuracy in excess of 13 to 14 bits. To yield higher performances, different approaches are necessary. The dual-slope technique and charge integration A to D converter considered in Chapter 6 offer improved accuracy but are slow. The Delta–Sigma converters considered in Chapter 7 offer the best compromise. Accuracy of 16 and more bits is obtained, regardless of the technology they are implemented in. Their high degree of accuracy is the result of a compromise exchanging magnitude for time resolution. They belong to a category of stochastic, rather than deterministic, converters.

Experimentation is essential to become fully acquainted with the converters performance. Theory is unable, generally, to quantitatively trace the impact of impairments due to the large amount of data that must be handled. Simulation provides a clean physical insight. However, it may be very costly when performed at the transistor level. Fortunately, thanks to simulation tools like MATLAB, there are means to perform simulations that do not require such excessive computation times. A set of MATLAB experiments is listed in the appendix concluding the book. A toolbox adapted for converters is described in the same appendix. The special functions it contains are exemplified in the first part of the appendix under 'Introduction' and examples are given of the potential of the tool. Understanding MATLAB statements is a prerequisite to taking full advantage of this tool.

Acknowledgments

Part of the material presented in the book was gathered while teaching at the Catholic University of Louvain, Louvain-la-Neuve, Belgium. The contribution of doctoral students is gratefully acknowledged, especially Bernard Ginetti, Benoit Macq and Alberto Viviani. The author is indebted to Prof. L. Morren for the sense of rigor and precision that he communicated to him. Some topics were developed while contributing to international courses, like Eurochip (Belgium), Europractice (LIRMM, Montpellier, France) and Iberchip in Latin America, as well as at the Institut Supérieur d'Electronique and the Ecole Nationale Supérieure des Télécommunications in Paris, the Institut Supérieur d'Electronique du Nord in Lille, the Universities of Genova and Cagliari in Italy, the Institut Charles Fabry, Université de Provence, in Marseille, France and the Edith Cowan University, in Perth, Australia. The contribution for suggesting appropriate rephrasing from Melissa McCreery, who reviewed the text, is gratefully acknowledged.

Tervuren   P.J.
April 2000

Contents

1 Terminology, specifications and evaluation techniques

    1.1   Resolution   1
    1.2   Ideal D to A and A to D converters   1
    1.3   Real D to A and A to D converters   3
            1.3.1   Linearity in D to A converters   5
            1.3.2   Linearity in A to D converters   8
    1.4   Testing the static characteristics   10
            1.4.1   Testing D to A converters   10
            1.4.2   Testing A to D converters   10
    1.5   Testing the dynamic performances of D to A and A to D converters   14
            1.5.1   The FFT test   14
            1.5.2   The code density test (C.D.T.)   19
    1.6   The concept of converter bandwidth   23

2 Scaled D to A converters

    2.1   Resistor scaling   24
            2.1.1   The R2R scaler   26
            2.1.2   Binary data selection   28
            2.1.3   Typical R2R architectures   32
            2.1.4   A resistor-less MOS converter R2R architecture   34
            2.1.5   Auto-calibration   36
    2.2   Unit-element scalers   38
            2.2.1   Statistical averaging and tolerances   38
            2.2.2   Capacitive scaling   41
            2.2.3   Transistor scaling   49

3 High resolution parallel D to A converters

3.1	Introduction		54
3.2	Current scaling using the dynamic current division principle		54
	3.2.1	Dynamic current division principle	54
	3.2.2	High resolution D to A converter based on the dynamic current division principle	55
3.3	Segment converters		59
	3.3.1	A typical current segment D to A converter	62
	3.3.2	A voltage segment D to A converter	64
	3.3.3	A 16 bit MOS segment D to A converter	65
	3.3.4	An improved version of the previous converter	67
	3.3.5	The current copier charge sharing problem	72

4 Feedback A to D converters

4.1	The need for a sample and hold circuit		76
4.2	Sampled data spectral considerations		80
4.3	A to D converters based on feedback loops		81
	4.3.1	The charge redistribution A to D converter	83
	4.3.2	A simple comparator	85
	4.3.3	Auto-correction	87
4.4	Codecs		90

5 Algorithmic A to D converters

5.1	The cyclic algorithm		94
	5.1.1	The Robertson plot	96
	5.1.2	Implementation	99
	5.1.3	Accuracy issues	102
5.2	The RSD algorithm		118
	5.2.1	The Robertson plot	119
	5.2.2	Floating point RSD converters	121
	5.2.3	Current mode RSD algorithmic converters	125

6 Rampfunction converters

6.1	The dual slope A to D converter	127
6.2	Improving the speed of rampfunction D to A converters	130
6.3	A charge rampfunction converter	131

7 Delta–Sigma converters

7.1	Quantization noise	135
7.2	The Signal-to-Noise Ratio	139
7.3	Increasing the SNR to improve resolution	139
7.4	A linear approximation of A to D Delta–Sigma converters	142
7.5	The generic Delta–Sigma A to D converter	146
7.6	Simple first order implementation of a Delta–Sigma A to D converter	147
7.7	More detailed analysis of Delta–Sigma converters operation	150
7.8	Non-linear aspects of Delta–Sigma converters	155
	7.8.1 The Limit-cycle	157
	7.8.2 Stability of noise shapers	157
	7.8.3 Idle tones	161
7.9	Discrete versus continuous loop filter implementations	161
7.10	Widening the bandwidth	163
	7.10.1 Synthesis of high order noise shapers	163
	7.10.2 Multi-bit versus single-bit noise shapers	168
	7.10.3 Cascaded (MASH) converters	170
7.11	Bandpass A to D Sigma–Delta converters	172
7.12	The decimator	175
7.13	D to A Delta–Sigma converters	178

8 Fast A to D converters

8.1	Flash converters	183
	8.1.1 Impact of comparator impairments	186
	8.1.2 Metastability, encoding errors and digital correction	190

8.2		Subranging, two-step, multi-step and pipelined A to D converters	192
	8.2.1	Subranging architectures	192
	8.2.2	Two-step converters	193
	8.2.3	Recycling converters	195
	8.2.4	Pipelined converters	195
	8.2.5	Impact of imperfections	196
	8.2.6	Correction strategies	198
	8.2.7	Examples of multistep and pipelined converters	200
	8.2.8	RSD multi-step converters	203
	8.2.9	Optimization of the S.H. switching sequence	209
8.3		Folding converters	209
	8.3.1	The folding principle	211
	8.3.2	Folding and interpolation	212
	8.3.3	Comparative study of folding versus flash converters	217

References 220

Appendix

INTRODUCTION:		Why simulations?	228
1		'Analog' and 'digital' representations	228
2		The input data	229
	2.1	Input data for A to D converters	229
	2.2	Input data for D to A converters	230
3		Available conversion functions	232
	3.1	D to A converters	232
	3.2	A to D converters	234
		Algorithmic converters	234
		Delta–Sigma converters	235
		multistep converters	235
4		Testing tools	236
		Visualization	236
		Static characteristics	236
		Dynamic characteristics	237

5	Examples		237
	-01-	Static characteristics of scaled and segment D to A converters	237
	-02-	Histograms and cumulative curves of INL and DNL	238
	-03-	Spectral analysis of scaled and segment D to A converters	238
	-04-	Static characteristics, INL and DNL of A to D converters	238
	-05-	Two-step A to D cyclic converter waveforms	239
	-06-	Spectral signature of A to D converters	239
	-07-	Code density test of A to D converters	239
	-08-	SNR versus magnitude of A to D converters	239
	-09-	Spectral content of noise shapers	240
	-10-	Magnitude of signals after integrators and quantizer	240
	-11-	SNR versus magnitude of noise shapers	240
	-12-	Second order accumulate-and-dump decimator	241
	-13-	Robertson plot	241

List of Converter Toolbox Functions   242

Index   257

Terminology, specifications and evaluation techniques

1.1 Resolution

The vast majority of converters feature a linear input–output characteristic that may be summarized as:

$$V \approx \left(\frac{b_1}{2^1} + \frac{b_2}{2^2} + \frac{b_3}{2^3} + \quad + \frac{b_N}{2^N}\right) \cdot V_{FS} \qquad (1.1)$$

The scalar V, left, portrays the analog continuous world, whereas the vector b, right, forms the discrete digital world. The variable V may be voltage, a signal from a sensor, or any other continuous variable. The finite core of the digital data is illustrated by the confined series expansion between brackets. The bs form a string of zeros and ones which defines the coded word representing the discrete counterpart of V, the ***N-bit*** binary fractional expansion of V divided by V_{FS}, where V_{FS} stands for the ***full scale*** (**F.S.**) range of V. The first and last bits, 'b_1' and 'b_N', are called respectively the ***most significant*** and ***least significant bits*** (**MSB** and **LSB**) according to their respective weights.

The number of bits N sets the number of discrete levels 2^N of the converter, the so-called converter ***resolution***, which determines the smallest step size $V_{FS}/2^N$ that can be discriminated. Since V_{FS} rarely exceeds a few volts, steps become very narrow once the resolution exceeds 13 bits. In a 16-bit converter with a 2 V F.S. range, the steps are only 30 µV high. Such fine granularity leads to severe challenges in integrated converter design.

1.2 Ideal D to A and A to D converters

The upper plot of Fig. 1.1 shows the ***transfer characteristic*** of an ideal D to A converter, where N equals three. The eight discrete input codes are plotted horizontally, whereas the lengths of the corresponding vertical segments portray the corresponding analog outputs. To enhance visibility, generally all the end-points are connected by a broken line, delineating a series of plateaus separated by steps amid the input codes. This line has no particular physical significance.

2 | Terminology, specifications and evaluation techniques

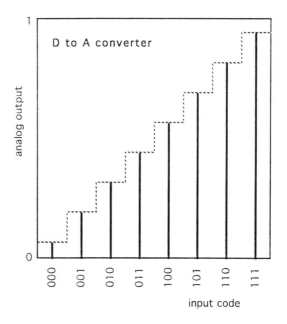

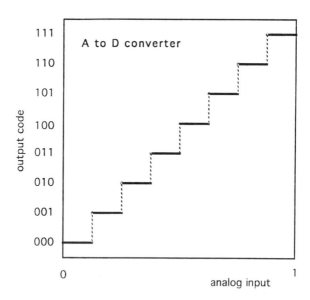

Fig. 1.1 Ideal transfer characteristics of a D to A (above) and an A to D (below) converter.

The lower plot of Fig. 1.1 shows the transfer characteristic of the ideal equal resolution A to D converter. The horizontal axis now represents the analog input whereas the vertical scale illustrates the finite set of digital output codes. The continuous character of the input data implies that all points along the horizontal axis have coded word correspondents. These change every time the input trespasses the so-called *transition* points, which portray the rounding of the output data to within ± one half *LSB*. Consequently, in A to D converters the lengths of the plateaus are significant, unlike in D to A converters.

1.3 Real D to A and A to D converters

Real converters diverge from their ideal counterparts by a number of impairments that must be apprehended to characterize their performances. To do this one must define appropriate evaluation criteria which hopefully lend themselves to easy evaluation. The target is not obvious since the appraisal may depend on the impact that the defects have on the overall performances of the device or system to which the converter belongs. For instance, how should one define linearity? In a measurement apparatus, it is a straightforward concept that implies the strict proportionality between the analog input and the digital coded output words. However this definition complies only partially for audio applications, since the ear is more sensitive to local perturbations than global distortion. Indeed, large differences between consecutive steps sound like clicks which generally produce severe annoyances, more so than large signal non-linear distortion. Thus, both global and local non-linearities should be discriminated. Another example is found in digital telecommunication systems. Here, the dynamic performances prevail over static for low harmonic distortion and intermodulation products are essential for the transmission quality. Consequently a unique definition encompassing all possible applications equally well is pure fiction.

This chapter reviews the main concepts used regarding D to A and A to D converter specifications and presents the experimental set-ups used for their performance evaluation.

Figure 1.2 shows the static characteristics of non-ideal D to A and A to D converters. The solid lines illustrate the non-ideal characteristics while the dashed lines reproduce the ideal characteristics shown in Fig. 1.1. In the upper D to A converter, impairments modify the heights of the plateaus while the steps' positions remain the same since, by definition, they are amid the input codes. In the lower A to D converter, the errors affect the transition positions while the heights of the plateaus remain unchanged, since these now represent the analog counterparts of the output codes. Both impairments have slightly different effects on the performance; consequently distinct measuring techniques are needed to assess linearity despite the fact that the objectives are the same.

4 | Terminology, specifications and evaluation techniques

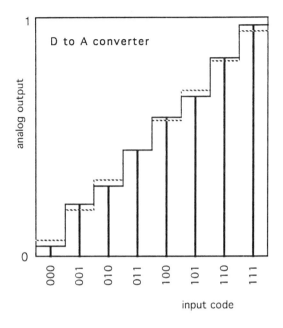

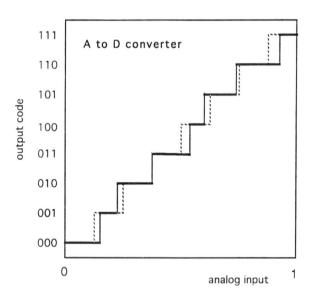

Fig. 1.2 Real transfer characteristics of D to A (above) and A to D (below) converters.

1.3.1 Linearity in D to A converters

Non-linearity in D to A converters affects the overall transfer characteristic as well as the step heights. The input–output characteristic shown in Fig. 1.3 is a rather unusual example where the heights of the steps experience little change but the overall transfer characteristic exhibits strong non-linearity. To describe this kind of non-linearity, one measures the difference between the actual transfer characteristic and some ideal characteristic—either the straight line joining the first and last points of the transfer characteristic (as in the figure) or the linear regression of the whole transfer curve. Regardless of the ideal characteristic, this difference is called the ***integral-non-linearity*** (or **INL**). To comply with the alleged accuracy of the converter, it should not exceed one-half LSB (Hoeschele 1994; Razavi 1995).

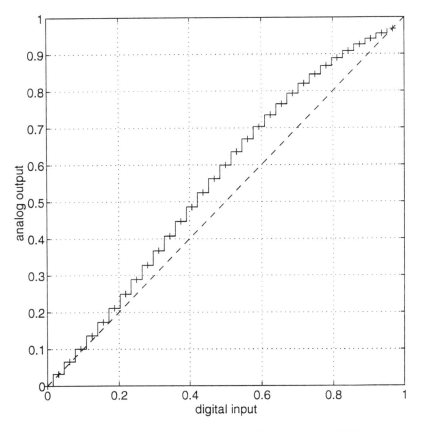

Fig. 1.3 D to A converter showing a strong Integral Non-Linearity (*INL*).

6 | Terminology, specifications and evaluation techniques

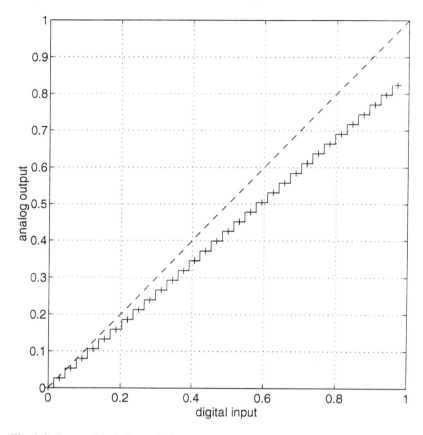

Fig. 1.4 The non-ideal slope of this otherwise perfect transfer characteristic should be identified as a gain error.

The *INL* should not be confused with the **gain error,** which characterizes divergence of the actual transfer characteristic slope from the ideal. In the plot of Fig. 1.4, the transfer characteristic is admittedly linear, but the slope does not coincide with the expected 45-degree input–output reference slope. The converter's gain is too small. This error is important in measurement equipment, but in audio it does not matter since volume control is customary.

The *offset error* describes the global shift of the transfer characteristic with respect to the ideal. Both the gain and offset error are derived from the linear regression of the transfer characteristic.

Figure 1.5 represents the transfer characteristic of a converter, which exhibits quasi-random and occasionally large differences between consecutive steps. Unlike the characteristic shown in Fig. 1.3, this transfer characteristic is quite common. The *differential-non-linearity* (or **DNL**) concept is introduced

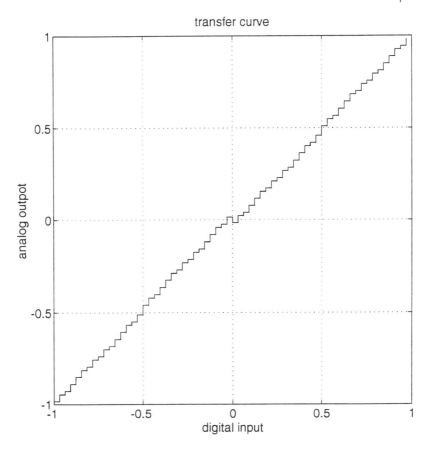

Fig. 1.5 Unequal heights of consecutive steps are measured by the Differential Non-Linearity (*DNL*). In the middle, severe *DNL* causes non-monotonicity.

to characterize this type of error. It portrays the irregularities between consecutive step heights and is evaluated in terms of LSBs minus one, since the *DNL* describes deviations with respect to the height of the ideal *LSB* step. The *DNL* is the discrete derivative of the *INL* error.

Negative steps like those in the middle of the graph of Fig. 1.5 are sometimes experienced. This large DNL error, called **non-monotonicity**, deserves special attention as it may cause serious problems in circuits where a converter closes a feedback loop. Non-monotonicity is similar to a local sign change of the transfer characteristic slope—the loop gain is positive instead of negative—jeopardizing stability. Monotonicity may be paraphrased as 'larger input obtains larger output'.

8 | Terminology, specifications and evaluation techniques

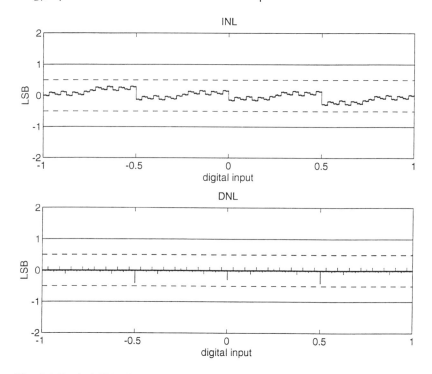

Fig. 1.6 Typical *INL* (above) and *DNL* (below) plots of a 10-bit D to A converter.

The input–output transfer characteristics considered so far are not the most efficient to appraise linearity. Steps are hard to see over 8 bits, and from 10 bits on the transfer curve looks more like a continuous line. The *INL* and *DNL* plots versus the input magnitude are more useful. The curves shown in Fig. 1.6 refer to a 10-bit D to A converter. The quasi-periodic pattern of the *INL* plot exhibits sharp steps that correspond to high order bit changes. The *DNL* curve also shows sharp peaks, corresponding to the *INL* discontinuities. Although one curve may be derived from the other, each one illustrates a feature not necessarily highlighted by the other. For instance, the *DNL* of the converter considered in Fig. 1.3 is satisfactory, but its *INL* well exceeds its specifications.

1.3.2 Linearity in A to D converters

Applying *INL* and *DNL* concepts to A to D converters requires a little reworking due to the opposed combinations of discrete and continuous data. The plot of Fig. 1.7 shows the input–output characteristic of a converter whose errors are intentionally exaggerated, in order to illustrate a few typical impairments.

As previously mentioned, the horizontal plateaus of the transfer characteristic are significant for they portray the slots over which the analog input may vary, without producing a change in the output codes. Ideally, all the plateaus should have the same lengths, but because of the converter's impairments some are longer, shorter, or vanish. First, the *midpoints* are defined, which correspond to the end-points in D to A converters. Ideally these points are equally spaced from every plateau transition. Taking the linear regression of the midpoints defines the converter's *gain* and *offset*. The *INL* is then the departure of the midpoints with respect to the gain line, expressed in *LSB* units as in D to A converters. The main difference is that it is evaluated

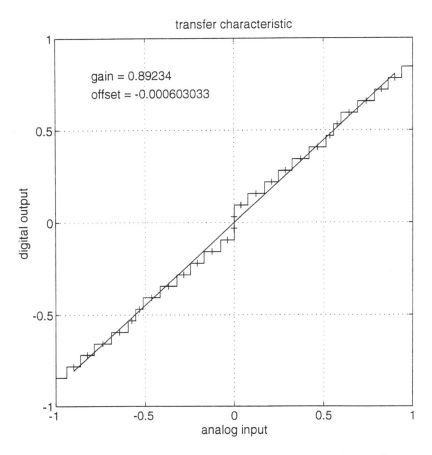

Fig. 1.7 Transfer characteristic of a non-ideal A to D converter. Gain and offset are derived from the midpoints linear regression.

horizontally instead of vertically. Similarly the *DNL* describes the different lengths of consecutive plateaus compared with the ideal length. The *DNL* is also expressed in terms of LSBs minus one and is like the discrete derivative of the *INL*. A new concept, called **missing codes,** is introduced, which relays the concept of non-monotonicity by reporting the presence of eventual zero length plateaus, such as those found in the middle and borders of the transfer characteristic shown in Fig. 1.7. Missing codes generate minus one errors in terms of DNL.

1.4 Testing the static characteristics

Experimental evaluation of the static transfer characteristic, as well as of the *INL* and *DNL*, requires very precise measurement equipment for resolution exceeding 10 bits. Since the amount of data increases exponentially with the number of bits, the measurement time easily exceeds acceptable limits. Thus, a dedicated automated test environment becomes essential.

1.4.1 Testing D to A converters

Generally an automated measurement stand consists of a high precision programmable reference voltage source and a computer. The computer generates the full set of code words applied to both the converter being tested and the precision reference D to A source. 'High precision' means that the resolution of the reference converter is at least 2 bits higher than that of the Device Under Test (D.U.T.). The data from the D.U.T are compared with those of the reference source. Care must be taken to avoid common mode problems associated with the comparison of large, almost equal, analog signals. The difficulty may be overcome if the data from the D.U.T. are reconverted into digital data by means of a high precision A to D digital voltmeter, and the comparison is made in the digital domain.

1.4.2 Testing A to D converters

Measuring the DC transfer characteristic of A to D converters requires a high precision analog reference source and a computer like that above cited. The latter is intended to compare the target codes with the output codes of the D.U.T. Testing high resolution converters is more subtle than D to A converters, because the noise from the Op Amps and comparators inside the converter often affect the transitions, causing jitter. As a result, near the transitions, the same input may generate adjacent code words repeatedly, especially since the resolution increases as the steps decrease. Averaging the output data over many tests leads to a probabilistic input–output characteristic like that shown in Fig. 1.8.

Testing the static characteristics | 11

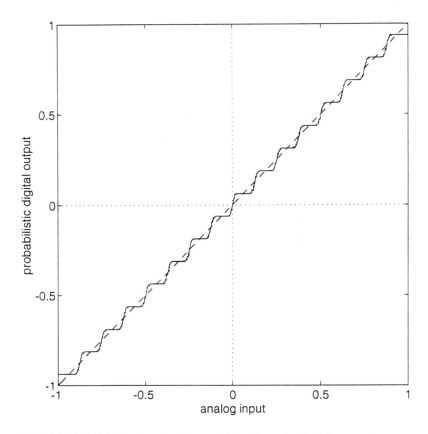

Fig. 1.8 Probabilistic input–output characteristic of a noisy A to D converter.

Under these circumstances, transitions are determined statistically. Two techniques presently used are described hereafter. Whichever is used, the time to complete a full transfer characteristic may be very long because many input samples are needed to define the transitions, and measurements must be repeated several times to determine the average noise of the transfer characteristic.

The ramp test

In the *ramp test* illustrated in Fig. 1.9, the computer generates a set of target codes that control a precision D to A converter. The target code sizes are at least 2 bits larger than those of the A to D converter to allow room for adjustment of the amplitude of the analog signal fed to the D.U.T. The transition acquisition algorithm is illustrated in the lower part of the figure.

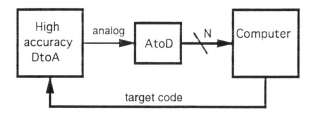

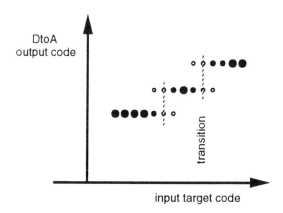

Fig. 1.9 Ramp test set-up and measurement strategy.

Circles symbolize the coded output words of the D.U.T versus the target codes. The sizes of the circles intuitively illustrate code word occurrences. Large circles characterize unambiguous code words whereas small circles represent occurrences shared by adjacent code words. When the circle sizes are identical, the codes exhibit equal occurrences. Any input that produces this situation is defined as a transition. Once all transitions are recorded, the computer determines the midpoints amid transitions and from this the gain, *INL* and *DNL* errors.

The servo loop test

The servo loop test shown in Fig. 1.10 was introduced by the Burr–Brown Company (Burr–Brown 1987). Its objective, as in the ramp test, is to determine the D.U.T. transition points. To reach this goal, the code words delivered by the converter are compared with the computer's target codes.

Testing the static characteristics | 13

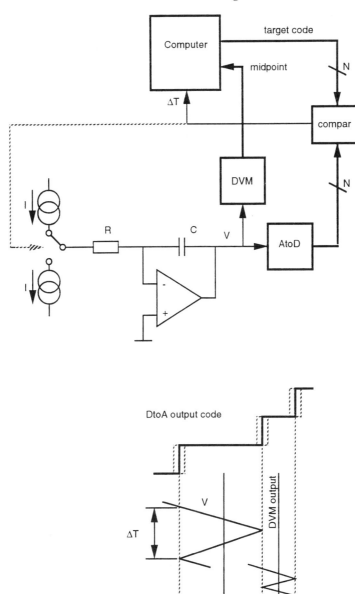

Fig. 1.10 The servo loop test set-up and measurement strategy.

When identical codes are sensed, the slope of the ramp voltage applied to the D.U.T. is reversed by changing the sign of the current fed to the integrator. As shown in the lower part of the figure, the voltage applied to the D.U.T. moves back and forth between adjacent transitions, one of which is the target code. While the servo loop is locked, the precision digital voltmeter determines the midpoint as the average of its readings. Once a midpoint is reached, the computer increases the target code to start a new evaluation procedure. Like in the ramp test, the servo loop test collects data that later enable the computer to determine the gain, *INL* and *DNL* errors. Note that a precision digital voltmeter is required instead of a high accuracy reference D to A.

1.5 Testing the dynamic performances of D to A and A to D converters

Static transfer characteristics contain useful data regarding D.C. performance, but no information about the dynamic behavior of the converters. Nonetheless, this is a very important item. For instance, resolution versus input signal frequency is a way to evaluate the converter bandwidth.

Deriving dynamic performances from static tests is clearly possible, but difficult to perform as precision reference sources are not fast devices. The time to smooth out switching transients limits the speed of the reference source. Most references cannot run at the speed the D.U.Ts are supposed to run. This may be overcome by running the reference source at a sub-multiple rate of the converter clock frequency, and choosing an input signal frequency slightly above or below the converter sampling frequency. Thus, the output follows the input signal at a rate which can be made arbitrarily slow. This technique, called the ***beat frequency technique,*** does not require that the reference source operate at the same rate as the D.U.T.

More appropriate techniques are described in the following. These are the Fast Fourier Transform (fft) and the Code Density Test (C.D.T.). The first applies to both types of converters, while the second applies only to A to D converters.

1.5.1 The fft test

In the fft test, the resolution of the converter is derived from the output signal *signal-to-noise ratio* (*SNR*). The basis of this test is to compare the relative magnitudes of the input signal spctrum, a high purity sine wave, and the noisy spectrum of the converter. Besides the fundamental ray that represents the input sine wave, the spectrum displays two distinct contributions: quantization noise and extra noise. The quantization noise is caused by the finite length of

Testing the dynamic performances of D to A and A to D converters | 15

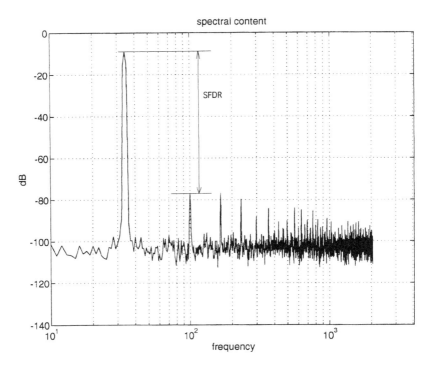

Fig. 1.11 The *fft* spectral signature unveils the dynamic counterparts of the *INL* and *DNL* by the emergence of harmonics and extra noise.

the coded data, regardless of whether D to A or A to D converters are considered. Quantization noise does not reflect defects; rather it is the consequence of the finite length of the digital output data. Other contributions reflect the converter impairments: *DNL*, *INL*, internal noise etc. These cause harmonic distortion and add extra noise to the quantization noise floor. Comparison of the ideal with the real spectra is the basis of the fft test.

The sensitivity of the fft test is very high (see Chapter 7 for the theoretical background). Figure 1.11 shows the fft of a 10-bit D to A converter output signal whose input consists of 33 entire sine wave periods. The large fundamental ray left illustrates the input sine wave. Some components of the converter display small mismatches, resulting in a 50% chance that the converter fail the one-half *LSB DNL* specification. The impact of the errors on the spectral power density is clearly shown by the large number of odd harmonics that are well above the background noise level. A simple way to characterize this is to estimate the difference in dB between the magnitude of the fundamental and largest harmonic ray. This difference, which in the example is 68 dB, is called the ***spurious free dynamic range*** (**SDFR**).

16 | Terminology, specifications and evaluation techniques

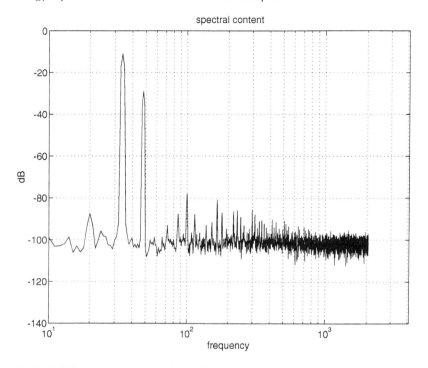

Fig. 1.12 When two sine waves instead of one are applied to a non-ideal converter, intermodulation products are generated.

The fft is not only very sensitive, it is also a very powerful analysis tool. In Fig. 1.12, the same converter as that shown in Fig. 1.11 is fed by two incommensurable pure sine waves whose peak magnitudes are respectively 0.8 and 0.1 V (the converter input dynamic range is limited to −1 and +1 V). The spectrum exhibits both harmonics and inter-modulation products. Their relative importance leads to the assessment of the third harmonic intercept that is currently used in telecom applications.

As previously cited, the fft provides a convenient method to determine the converters' resolution and dynamic range, based on the following expression, demonstrated in Chapter 7.

$$SNR_{dB} = 6.02N + 1.76 \qquad (1.2)$$

The equation establishes a clear link between number of bits N and SNR. Deriving the SNR from the fft, determines the resolution. To evaluate the SNR, we measure the power of the input sine wave and compare it with the total noise power. The input signal power stems from the fundamental ray and the noise power from the noise density integrated over the frequency band from

Testing the dynamic performances of D to A and A to D converters | 17

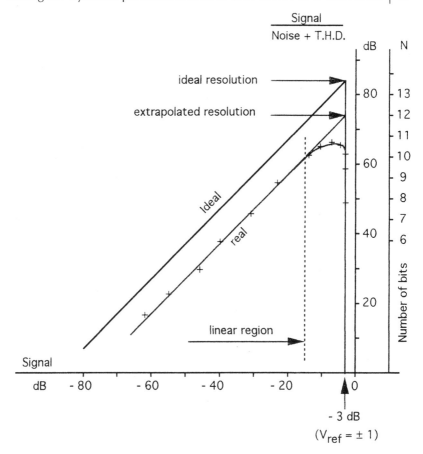

Fig. 1.13 The signal-to-noise ratio of the output versus the magnitude of the input signal provides a means to evaluate the resolution and dynamic range.

minus to plus one-half the sampling frequency. Dividing the first by the second yields the *SNR*.

The above equation is only valid as long as the magnitude of the input sine wave encompasses exactly the F.S. range of the converter. Unfortunately, this condition often enhances the distortion of the converter since saturation tends to increase the harmonic content near the F.S. This may then lead to the inconsistent conclusion that the resolution derived from eqn (1.2) is smaller than it really is. In fact, the resolution is related to the step size, the smallest increment of the converter's output: a concept that obviously ignores large signal distortion. To avoid this, the resolution evaluation procedure is generally the following. The magnitude of the input sine wave is decreased until 40 to 80 dB below full scale. The *SNR*s are then plotted versus the growing input signal power. In an ideal converter, the result should be a straight line such as the thick line shown in Fig. 1.13. The real converter deviates from the ideal in two ways. When the magnitudes are small, the points align along a straight line parallel, though slightly below, to the ideal *SNR* line, experiencing the resolution loss from converter impairments like DNL. Second, when the magnitude of the input signal increases, the *SNR*s tend to level off due to the growing harmonic distortion. Extrapolating the straight line joining the small to the full-scale signal, the *SNR* is evaluated regardless of the large signal distortion. For example, if the dynamic range spans −1 to +1, the full-scale signal power corresponds to −3 dB (the rms value of the input is one over the square root of two). The point where the extrapolated *SNR* curve intersects the vertical line corresponding to −3 dB fulfils the condition that justifies using eqn (1.2). The ***effective number of bits* (ENOB)** is then determined by solving eqn (1.2) with respect to N. Thus:

$$ENOB = \frac{SNR_{dB} - 1.76}{6.02} \qquad (1.3)$$

The actual *SNR* curve versus the input signal magnitude is generally referred to as the '***signal to noise-plus-total-harmonic-distortion***' or the ***S/(N + THD)***.

Let us consider an example: suppose that the extrapolated *SNR* of an alleged 10-bit converter reaches 55 dB instead of the ideal 62 dB predicted by eqn (1.2). The *ENOB* according to eqn (1.3) is thus 8.84 bits. The resolution of this converter totals 8 bits instead of 10.

To be effective, the fft requires some care to avoid possible misinterpretations. Indeed, the Fast Fourier Transform is prone to spectral leakage, a phenomenon occurring whenever discontinuities appear at both ends of the observation sample. When this is the case, the peak corresponding to the fundamental tends to smear, disturbing the spectral distribution. The problem is avoided if an integer number of periods is chosen and the data prior to the fft computation are windowed in the same time. A commonly used window is the Hanning window (Harris 1978), which unfortunately widens the fundamental spectral ray in the same time. This is not a problem, as long as all the points belonging to the fundamental ray are included in the power count. Another problem encountered is aliasing (folding back) of harmonics within the converter frequency band.

The *SNR* test is applicable to D to A as well as to A to D converters, but the set-ups are distinct. In the D to A test, the input is a perfect digital quantized sine wave. The analog output spectrum is evaluated by means of a spectrum analyzer. The limited dynamic range of the spectrum analyzer often requires a notch filter to remove the fundamental. The non-linear distortion is also often evaluated using a separate distortion analyzer. The test may be carried out in the digital domain, computing the fft of the output signal once it has been reconfigured into a digital signal by means of a high resolution and fast

Testing the dynamic performances of D to A and A to D converters | 19

auxiliary D to A converter. Precision means that the performances of the auxiliary converter surpass those of the D.U.T. by at least two bits. When the *SNR* test is applied to A to D converters, the analog input must be a low distortion sine wave and the spectral assessment done in the digital domain.

An improved version of the fft test has been proposed in (Boser 1988). In this approach, the separation between fundamental and noise is carried out in the time-, rather than in the frequency-domain. The fundamental of the output signal is approximated using a pure sine wave whose amplitude and phase are adjusted to obtain the best possible fit. What remains after subtracting the fitted sine wave from the output signal is noise. Otherwise, the evaluation proceeds as previously cited.

Since the *SNR* evaluation is a dynamic test, converters are put in a situation much closer to reality than in any static test. This offers the possibility to test the performances of the converter versus the frequency of the input signal.

1.5.2 The code density test (C.D.T.)

In the code density test, the converter performances are inferred from the histogram of the output code words (Doernberg 1984). The input is a known periodic signal spanning the full converter scale. The idea may be summarized as follows: slowly varying input signals produce many identical code words whereas rapidly changing signals produce very few. If a large number of samples are taken, the histogram of the output data tends to become a discrete representation of the input signal probability density. Since the input signal probability density is known a priori, all that is needed is to compare the result of the code density test with the probability density of the input signal. From this, one may infer the impairments of the converter. The technique is very efficient with regard to *DNL* errors.

The input signal is generally a high purity sine wave whose probability density $p(V)$ is given the following expression, where A is the peak magnitude of the sine wave:

$$p(V) = \frac{1}{\pi\sqrt{A^2 - V^2}} \qquad (1.4)$$

The histogram shown in the upper part of Fig. 1.14 illustrates the code density test of a perfect converter. As expected, there is almost no difference between the staircase shaped histogram and the sine wave probability density, illustrated by the continuous curve. In practice, many samples are needed to attain a satisfactory degree of confidence of the final result. For instance, the number of counts per bin required to yield a precision of 0.1% with 99% confidence is one thousand according to (Doernberg 1984). For a 12-bit converter, the total number of samples would thus be 4 096 000 counts.

20 | Terminology, specifications and evaluation techniques

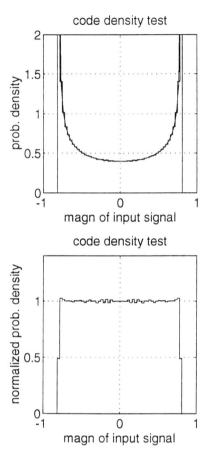

Fig. 1.14 The upper graph shows the CDT (Code Density Test) of an ideal 8-bit A to D converter, the lower graph illustrates the ratio of the Code Density Test over the ideal sine wave probability density whose peak magnitude equals 0.8 V.

Considering a converter sampled at 8 kHz, this represents a total measurement time of 8.5 minutes. The plot below shows the histogram after it has been divided by the sine wave probability density. A horizontal line is obtained when the converter is ideal. When the lengths of the plateaus are either too long or too small, the histogram senses the difference with respect to the ideal. The result is similar to a *DNL* plot. The histogram of Fig. 1.15 was obtained after a small source of non-linearity was introduced in the previous converter. Although the shape of the histogram remains roughly the same, some code words are seriously corrupted. In the middle, missing codes are spotted and a few other code words seriously degraded. The curve below underlines the fact that the lack of occurrences of some code words is compensated by a small

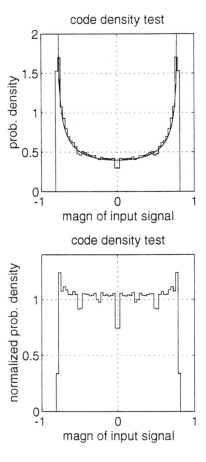

Fig. 1.15 The Code Density Test offers a simple means to evaluate the *DNL*.

shift upwards of the entire histogram. The rapid increase near the borders is, in fact, an artifact caused by the converter gain error. Since the gain is not unity, the magnitudes of the input and the output sine waves no longer coincide. If they were adjusted, the artifact would disappear. The conclusion is that the interpretation of code density tests requires some caution. Slow changes of the magnitude of the input signal during the C.D.T. are namely a potential source of problems that affect the histogram adversely.

Among the conditions that must prevail in order to avoid artifacts, two are very important: first, the input signal should always consists of an integer number of periods to maintain the probability density equal to the anticipated one; and second, the number of sine wave periods and the number of samples should never be multiples of each other, to maintain the chances of occurrences of the code words identical over the entire F.S.

Noise has an unexpected effect on the validity of the code density test. In high accuracy converters, for example over 13 bits, the step size is so small, especially in low voltage converters, that internal wide band noise may be of the magnitude as the steps. The sum of the input signal and the noise then encompass two or more code words at once so that a missing code may be overlooked. If the number of samples is very large, some missing codes may even fade out from the tally. Under these circumstances, a converter could comply with its alleged specifications, despite missing codes—a surprising but true fact (Ginetti and Jespers 1991).

Despite these problems, the code density test offers a number of interesting features. It detects missing code, evaluates the gain and detects errors, offset, *INL* and especially *DNL* errors. The test does not require a high precision programmable reference source, but only a high purity sine wave generator. Besides its simplicity, it is a dynamic test like the fft. However, it can only be used for A to D converters.

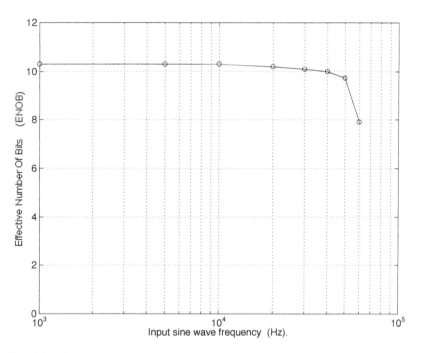

Fig. 1.16 The converter bandwidth can be derived from the plot representing the *ENOB* versus frequency.

1.6 The concept of converter bandwidth

The time needed to process the data inside converters generally fixes the maximum allowable sampling rate. This rate is usually expressed as kilo-Samples (kS/s) or Mega-Samples per second (MS/s). Any attempt to increase the sampling rate beyond this limit inevitably lowers the performance. The boundary between what is and is not acceptable defines the so-called *converter bandwidth* (Hz).

Evaluation of the converter bandwidth generally relies on the evaluation of the *ENOB* versus the input signal frequency. Another test is to maintain the constant input signal frequency, while changing the sampling rate. The first case is illustrated in Fig. 1.16. Once the *ENOB* drops by 1/2 *LSB* with respect to the maximum resolution, the alleged accuracy is lost. In the example of Fig. 1.16, the 10-bit accuracy is lost above 50 kHz. The method is applicable to both D to A and A to D converters.

Scaled D to A converters   2

In this chapter we review a set of converters that output data constructed by adding analog reference weights to each other. The references form a source **scale**, consisting of calibrated current sources, voltage sources or charges, whose selection occurs under control of the input words.

Scaled converters are fast by essence. What determines their conversion speed is the time needed to decode the input word, select appropriate weights and sum them. These operations can be done in a single clock time, consequently bandwidths as large as 500 MHz are feasible. Generally scaled converters do not exceed ten bits resolution, due to mismatches in the reference scales. Higher resolutions require laser trimming or self-calibration techniques.

Three kinds of converters are reviewed in this chapter according to the kind of weights they use, resistive-, capacitive- and transistor-scaled converters.

2.1 Resistor scaling

First, discrete resistors were used to implement reference scales in D to A converters. With the trend to put more devices on-board the chips, the need to integrate the resistors also became evident. Integrated 'precision' resistances can be fabricated by etching conductive stripes in thin layers of resistive material. What determines their magnitude is the length over width ratio and the resistance per square of the layer in which they are implemented. Generally the layers consist of NiCr or aluminum deposited on field oxide. Poly-silicon layers can also be used. Junctions are preferably avoided for their pronounced non-linear behavior due to the underlying junctions.

The absolute accuracy of integrated resistances is generally very poor but matching (Gray and Meyer 1993) is practicable as illustrated in Table 2.1. The accuracy is a compromise between tolerances and area. The etching process used to delineate the resistive stripes is subject indeed to under-etching, causing random width variations whose magnitudes should be small compared with the widths of the resistors. Thus, wide resistors feature matching figures that are necessarily better than narrow resistors, with a trade-off however between precision and area. It is generally admitted that 0.2 or 0.1% tolerances represent a reasonable compromise. Better figures require overly large devices.

Table 2.1
Tolerances of various types of integrated resistiors (Gray et al. 1993)

Resistor type	Sheet resistance (Ω/square)	Absolute tolerance (%)	Matching tolerance (%)	Temperature coefficient (ppm/°C)
Base diffusion	100–200	± 20	± 2 (5 µm wide) ± 0.2 (50 µm)	+ 1500 to 2000
Emitter diffusion	2–10	± 20	± 2 ± 1 (5 µm wide)	+ 600
Ion implanted	100–1000	± 3	± 0.1 (50 µm)	± 100
Epitaxial	2000–5000	± 30	± 5	+ 3000
Thin film	100–2000	± 5 ... ± 20	± 2 ... ± 0.2	± 10 ... ± 200

Layout plays an important role to ascertain matching. Bends and twists, inevitable when implementing large resistances, do not impair matching too much. Interconnect pads are more awkward because they are likely to introduce unwanted voltage drops, especially when low-valued resistors are being implemented (10 ohms or less). The best way to prevent this is to take advantage of voltage sensing by using separate current-less pads (Wittmann 1995). Another important source of mismatch is the eventual non-uniformity of the sheet resistance. Concentration gradients are inherent to any fabrication process. Even when small, they create a changing mismatch bias over space that must be counteracted. Centroid layouts are advocated to neutralize first-order gradients. Another possibility is to address all resistances accordingly to a pseudo-random order in space.

2.1.1 The R2R scaler

A simple resistive D to A converter can be implemented as a string of binary weighted resistors connected to the summing node of an Op Amp like that shown in Fig. 2.1. The other end of each resistor is connected to either a reference voltage or the ground, depending on the bit it unfolds. This is the worst possible candidate for integration as it requires a wide range of resistances. The least significant bit is implemented by means of a resistor 2^N times larger than that defining the most significant bit. Since the latter is the most stringent as far as accuracy, its size must be large. Such a reference scale is only acceptable when the number of bits does not exceed 2 or 3.

A better approach is shown in Fig. 2.2. It requires only two distinct resistances so that the same unit-resistor can be used everywhere; factors 2 or $\frac{1}{2}$ being implemented by means of series or parallel combinations. Consequently matching is feasible.

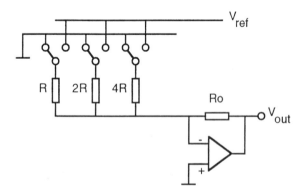

Fig. 2.1 Scaled resistive converter making use of binary weighted resistors. This type of converter is not suitable for integration above 3 bits.

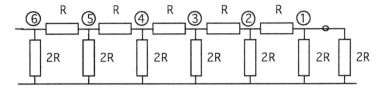

Fig. 2.2 R2R ladder networks take advantage of matching because the same resistor can be used to implement the horizontal and vertical legs of the network.

When a voltage is applied to node 6 in the circuit shown in Fig. 2.2, a binary voltage scale builds up along the upper nodes. The impedance between node 1 and ground is equivalent to a single resistance R. Thus, the series combination of all the resistors to the right of node 2 is a divide-by-two network. This is also true for node 3 with respect to node 2, and so on. Thus, all voltages along the upper nodes implement a binary scale. The same applies to the currents flowing in the vertical resistances $2R$.

As previously cited, thin layers of NiCr achieve at best 0.1 per cent accuracy without trimming. To evaluate the accuracy achievable with R2R converters, the results of a Monte Carlo simulation are presented in the following. The simulation does not take into account any perturbation other than resistance mismatch, thus no consideration is made of sheet resistivity gradient, non-linearity or temperature dependence. Mismatch affects the *INL* and *DNL* all over the converter's dynamic range but preferably at mid-code transitions, every time words or sections of words switch from a zero followed by a long string of ones to a one followed by a string of zeros. This is clearly illustrated in the plots of Fig. 2.3, which display the *INL* and *DNL* of five 10-bit R2R converters whose resistors are supposed to obey a Gaussian distribution with a standard deviation '*sigm*' equal to 0.5×10^{-3}. Sharp errors are clearly spotted when the MSB's change, especially in the middle. Notice that the *DNL* is twice as large as the *INL* in the middle, since the steps of the transfer function unfold symmetrically around the converter's linear regression.

The two histograms displayed in the upper plot of Fig. 2.4 are taken from a sample of five thousand R2R 10-bit converters whose resistances conform identical Gaussian error distributions and whose standard deviation equal 0.5×10^{-3}. The curves relate to the largest *INL* and *DNL* figures determined in every transfer characteristic. The cumulative histograms shown in the lower plot of Fig. 2.4, which relate to standard deviations equal to 0.5×10^{-3}, 1×10^{-3} and 2×10^{-3}, answer the question: how many converters meet the one-half *LSB* tolerances? Respectively 92, 58 and 18% of the converters conform to the one-half *LSB* tolerances as far as *DNL* and 100, 87 and 43% as far as *INL*. Tolerances as low as 0.05% afford excellent yield but are hard to achieve without a substantial increase of area or the use of laser trimming.

28 | Scaled D to A converters

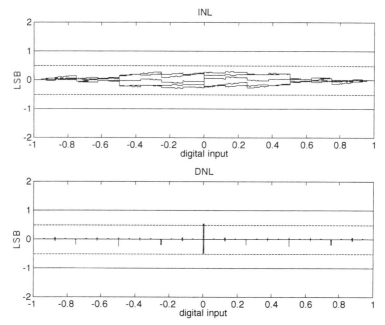

Fig. 2.3 *INL* and *DNL* plots of five 10-bit R2R converters implemented with 0.05% matched resistors.

With 0.1% tolerances, a reasonable number of converters lie within their specifications, but the number of rejected devices rapidly increases when the tolerances surpass 0.2%. Hence, 10-bit resolution R2R converters are feasible, but above 10 bits, laser adjustments become essential. The laser burns small perpendicular slits in every resistor so as to increase slightly the magnitude of the resistance. However the ultimate achievable adjustment does not exceed 12 bits because laser trimming is a brutal thermal treatment. Every step induces thermal stresses, as reflected by previously trimmed resistors.

2.1.2 Binary data selection

How are binary weights selected and added to obtain data uncorrupted by other sources of inaccuracy like switch impairments, which are likely to introduce voltage pedestals, leakage currents and transients during switching?

Most R2R converters rely on current sensing. Switching is achieved through bipolar or MOS transistors. Bipolar transistors are poor voltage switches but good current switches. A saturated bipolar transistor is like a switch with a voltage pedestal in series that is more or less 100 mV large regardless of the collector current that flows in the transistor. When turned

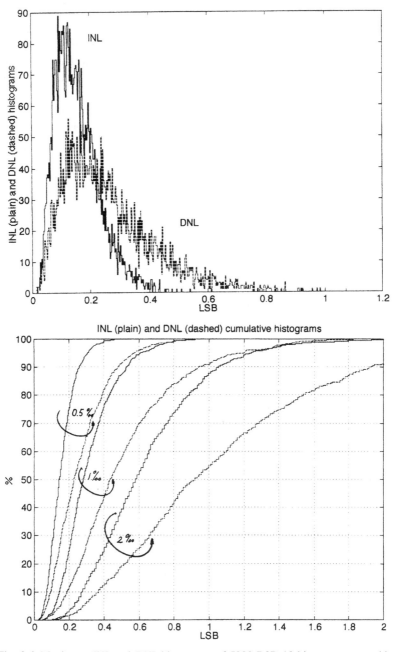

Fig. 2.4 Maximum *INL* and *DNL* histograms of 5000 R2R 10-bit converters making use of 0.05% matched resistors (above). Cumulative histograms of the same converters making use of 0.05, 0.1 and 0.2% matched resistors (below).

upside down, the pedestal is smaller but the gain of the transistor is very small. Bipolar transistors do much better when used as current switches, such as in ECL-gates, and their leakage current is no larger than 1 to 10 pA at room temperature. MOS transistors lend themselves to the implementation of both types of switches. In the so-called linear region, they behave like voltage switches in series with small linear resistors. When cut-off, they are somewhat inferior to bipolar transistors because their leakage current is attributable not only to junction leakage but also to subthreshold current, which is highly temperature sensitive.

In the MOS circuit represented in Fig. 2.5, the current in the vertical 2R resistor is sensed by a double MOS voltage switch. The current flows either to the ground through transistor Q_{1a} or to the summing junction of the Op Amp through transistor Q_{1b}. If the sizes of the two transistors are the same, the voltage drop across the switch remains constant whichever outcome. However the ensuing voltage drop across the switch slightly reduces the voltage across the resistor 2R, and thus the current. To maintain the binary scale unharmed, all switches should withstand the same voltage drop. The best way to do this is to scale the W/L ratios to equal the delivered currents that they transfer. Thus, every transistor pair width is to be multiplied by two when going from the least to the most significant bit.

In bipolar circuits, switches are always the ECL-type circuits as shown in Fig. 2.6. A common base transistor separates the switch from the upside-down ladder network below. This transistor is necessary to keep the voltage across

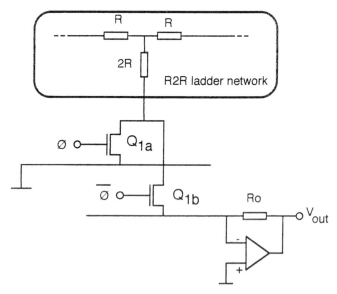

Fig. 2.5 A MOS transistor read-out circuit voltage switch.

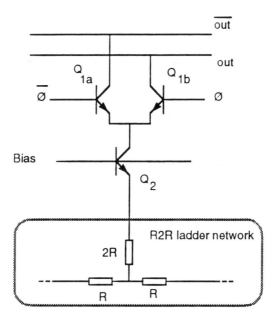

Fig. 2.6 A bipolar transistor read-out circuit current switch. Transistor Q2 isolates the R2R network from the switch.

the ladder network away from voltage variations experienced at the emitters of Q_{1a} and Q_{1b}. These transistors operate like source followers: one is 'on', the other 'off'. The common emitter voltage follows the largest of the two logic signals controlling the switch. Q2 affords better control of the DC potential applied to the ladder network since the line called 'bias' fixes the potential of all the vertical resistors. Perturbations experienced at the collector of Q_2 are divided by the intrinsic gain of this transistor, which may be as large as a few hundred. Since the voltage drop across the base-emitter space is a function of the current flowing in each transistor, the emitter sizes must be scaled like the W/L ratios above.

Switches, whether MOS or bipolar, are prone to transients proceeding from the clock signals. They are the outcome of time-skew, substrate noise, parasitic capacitances or charges, etc. They look like pulses currently designated **glitches** whose magnitude is evaluated in terms of Volt.second, or power spectral density, and sometimes peak amplitude. For example, consider the bank of switches controlling the reference weights in any of the two previous circuits. All switches are supposed to operate synchronously but, in practice, the waveforms that control them may be softened and may not be strictly synchronous. Differential capacitive loading may be the cause. When this happens, a situation is created during which the code word applied to the reference bank briefly assumes invalid states. As a result, the converter tries to reach conflicting values during a very short time. This produces a kind of spurious pulse that is superposed on the output. Another major impairment regarding the switches is charge injection. In MOS transistors, the charge related to the overlap capacitance and inversion layer cannot simply vanish, it must go somewhere. When injected in a capacitor it produces a small voltage step, which introduces an offset. Charge injection is a major problem that is dealt with more thoroughly in later chapters.

Because glitches impair the spectral signature of converters, their impact must be reduced as much as possible. Time skew can be lessened using a register that waits for all transients to die out before transferring code words to the reference bank. Charge injection is counteracted in another way. In well-designed switching networks, charge injection produces common mode perturbations that are ignored by the analog differential architecture. Examples will be considered further.

Having considered the switches, let us now examine how the selected currents are added. The simplest manner is to use current switches like those shown in Fig. 2.6 to inject the currents in a resistor. The high output impedance of the current generators prevents the output signal from reacting on the individual currents. The main advantage of this is the potential for large bandwidth it allows, for no active devices like Op Amps are involved in the summation. The other approach is to use an Op Amp like that shown in Fig. 2.5. The interaction between output and input data is avoided since the summing node behaves like a short-circuit. Essential items are now the gain-bandwidth product and slew-rate. The first determines the speed while the second affects the switching time by outputting signals whose slope remains constant regardless of the step magnitude. Another important issue is doublets or narrow pole-zero pairs located below the transition frequency of the Op Amp. Such doublets lengthen the settling time substantially once the Op Amp nears steady state conditions (Kamath 1974). Feed-forward architectures are prone to this kind of impairment.

2.1.3 Typical R2R architectures

A few typical R2R implementations are now considered. The first shown in Fig. 2.7 is a converter with an alleged resolution of 12 bits. A 7-bit wide R2R ladder network is easily visible in the middle. The two remaining MSBs make use of binary weighted resistances whereas the three LSBs on the right are implemented by means of scaled transistors. The feedback network on the left controls the base voltage of Qa and forces the current in this transistor to equal V_{ref}/R_o. The base voltage of Qa plays the same role as the bias line shown in Fig. 2.6. It is supposed to control the voltage all along the ladder network. But one must take into account the fact that each time current is multiplied by two,

Resistor scaling | 33

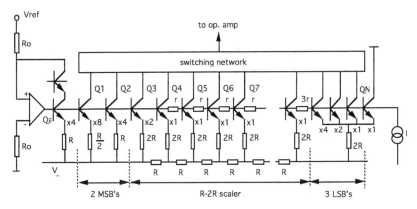

Fig. 2.7 A composite bipolar 12-bit converter making use of an R2R network (middle 7 bits), resistor scaling (2 MSBs) and transistor scaling (3 *LSBs*).

the base-to-emitter voltage of the correspondent transistor increases by 18 mV, accordingly to the expression $(U_T).\log_{10}(2)$ where U_T is the thermal voltage kT/q, equal to 26 mV at room temperature. Thus, to avoid a possible emitter's voltage skew, a similar but opposed skew must be imposed to the bases. To this effect, either small compensation voltage sources can be placed in series along the common base bias line, or one may maintain the base-to-emitter voltage constant by multiplying the emitter areas by two like the *W/L* ratios in the MOS circuit above. The converter represented in Fig. 2.7 shows the first option. Small identical resistors are placed all along the line feeding the bases and current is injected along the bias line to produce 18 mV voltage drops across each resistor. Different strategies are used for the *MSB* and *LSB* sections. In the *MSB* section, the emitter areas increase like the reciprocal of the feeding resistances to keep the base-emitter voltages constant. And in the *LSB* section, emitter scaling is used to divide the current from the rightmost resistor accordingly to binary weights. Connecting identical transistors in parallel does this. Transistor scaling will be discussed in more detail in Section 2.2.3. At the time this converter was designed, the general belief was that good performances could only be achieved with R2R architectures. We will see that transistor scaling offers a better approach.

The D to A converter shown in Fig. 2.8 exploits the voltage mode instead of the current mode readout (Guy and Trythall 1982). The output is the voltage sensed by the unity gain buffer. Equal currents must be injected into all the nodes of the R2R. Since they are identical, no compensation is needed for the buffer transistors. However accuracy is a function not only of the scaler but also of the current sources matching.

Modern converters no longer exploit R2R ladder networks despite their simplicity. Today, better performances are obtained with circuits that exploit

34 | Scaled D to A converters

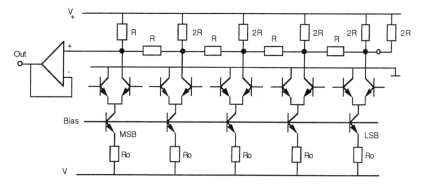

Fig. 2.8 Voltage mode read-out of an R2R converter. The mismatches of the current sources and resistances add up.

the paralleling of many unit- elements. This does not preclude resistors from being widely used even today. High speed converters, like flash, pipelined and multi-step converters offer typical examples later described in Chapter 8.

2.1.4 A resistor-less MOS converter R2R architecture

This section deals with a more recent circuit derived from the R2R network. It may be a good candidate for very low power converters. The circuit makes exclusive use of MOS transistors, no resistors are used; a rather surprising concept since linearity is not the main asset of MOS transistors! The idea is based on the fact that all that the successive cells are supposed to do in R2R ladder networks is to divide current into equal parts. Identical devices in parallel perform this operation, *whether linear or not*. Thus two identical MOS transistors in parallel, saturated or not, operating in strong or weak inversion, split current in two equal parts, just like resistors. The MOS converter based on this concept is shown in Fig. 2.9.

All the transistors in this circuit are supposedly identical and driven by the same gate voltage. Thus, advantage is taken of matching. The two cascaded transistors on the right are equivalent to a single device with one-half their original *W/L*. A transistor is shown on the left, placed in series with a switch that dumps current to the ground or to the summing node of the Op Amp. Whichever its state, the switch is equivalent to a single series transistor. Thus the series combination is nothing but a replica of the first branch. As a result the current from Q_R is equally divided between the two branches. Since these are in parallel, they are equivalent to a single transistor with the nominal *W/L*. Thus the series combination of Q_R and the latter is equivalent to the vertical leg of the next cell. Repeated cell after cell, this reasoning leads to the conclusion that all the currents in the vertical branches implement a true binary

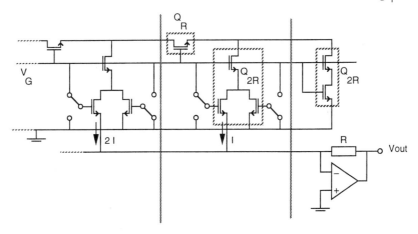

Fig. 2.9 In this all-MOS converter, current scaling emulates R2R converters but without resistors.

scale, like in R2R networks. The vertical branches stand for the resistors 2R, and transistors Q_R represent the series resistors R. Note that the voltages do not implement a binary scale. Note also that since the switches belong to the so-called 2R branches, they fulfil a double function: they are part of the vertical 'resistors' and they perform switching at the same time. In the old R2R version, resistors and switching transistors were distinct devices and the sizes of the switch transistors were critical in the accuracy of the ladder network. There is nothing similar in the all-MOS converter.

The current division principle has a broader base as shown in Bult (1992) and van den Elshout (1994). Consider the current divider illustrated by Fig. 2.10, which represents a single MOS transistor having an access somewhere along its channel. Current injected or extracted from this node is divided according to a ratio that is independent of the magnitude of the input current and of the voltages applied to the transistors. Similarly this ratio is not affected by the operating conditions of the MOS transistor, whether saturated or not, operating in strong, moderate or weak inversion. Consequently:

$$\frac{\Delta I_{Da}}{\left(\frac{W}{L}\right)_a} = -\frac{\Delta I_{Db}}{\left(\frac{W}{L}\right)_b} = \text{constant} \tag{2.1}$$

ΔI_{Da} and ΔI_{Db} represent the current variations resulting from extracted or injected current. The constant is a complicated function of the circuit parameters (Bult 1992) that takes into account a number of factors like drift and diffusion currents, as well as the vertical field dependent mobility

36 | Scaled D to A converters

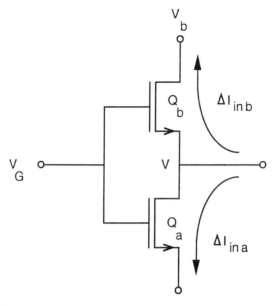

Fig. 2.10 Illustration of the current division circuit described in Bult and Geelen (1992).

(Tsividis 1987). The most important item influencing the accuracy is velocity saturation. Fortunately, in the all-MOS converter, all transistors operate in the linear regime. Thus velocity saturation is a lesser problem; all the longer channels further reduce its impact. The current division principle was confirmed experimentally by Bult (1992) who implemented an all-MOS attenuator with −80 dB total harmonic distortion, clearly illustrating the potential of all-MOS converters. Almost the same figures are reported in Hammerschmied and Quiuting (1998), which describes a 10-bit, 12 mW converter using transistors with channel lengths of 30 μm. The drawback of such large channel lengths is the poor sampling rate of this converter that reaches only 200 kS/s.

2.1.5 Auto-calibration

R2R scalers do not usually exceed 10-bit resolution, whichever technology and architecture are used. Self-calibration achieves better figures. In self-calibrating converters, the output is compared with a set of references during a calibration phase when correction data are computed and stored in an on-board memory. This occurs while the converter is idle. It is an occasional procedure that is only repeated when the converter is turned on. The circuit of Maio

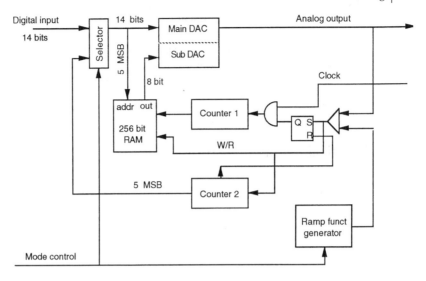

Fig. 2.11 An auto-correct 14-bit R2R converter (Maio *et al.* 1981).

(1981) shown in Fig. 2.11 is an early but interesting example of a self-calibrating R2R converter. The device consists of a main high-resolution converter supplemented by a low-resolution sub-converter to correct the errors from the main device. The main converter is an untrimmed 14-bit R2R network barely reaching 9 or 8 bits and the sub-converter is an R2R converter with 8 bit resolution. In normal mode, the 5 most significant bits of the 14-bit input words address a Random Access Memory (RAM) whose output controls the sub-converter. And this current is added to the current delivered by the main ladder network to compensate for its errors.

The question arises: How does the RAM get the appropriate data and what determines arbitration? The absolute reference is a rampfunction delivered by a Miller integrator whose linearity is supposed to be 18 bits. During self-calibration, the subconverter is inactive while the main converter is controlled by words generated on-chip. These words consist of two fields: a 5-*MSB* long word and zeros for the rest. Counter #2 to exemplify all possible codes from 1 to 32 generates the five *MSB*s. To begin, the rampfunction and counter are reset. Then Counter #2 is incremented and the rampfunction launched. The main converter delivers the first output of a sequence of 32 equally spaced DC levels unfolding the entire dynamic range. When the comparator acknowledges equal converter and rampfunction outputs, the flip–flop is set and counter #2 incremented. This freezes the clock count of Counter #1 and forces the main converter to output the next coarse level. If the main converter were ideal, all clock counts should be the same. The differences measure the transfer characteristic non-linearity. They are being used to compute the correction codes to store in the on-board RAM memory when driving the sub-converter in normal mode. Another example of an auto-calibrated R2R converter is described in Burr-Brown (1986).

2.2 Unit-element scalers

2.2.1 Statistical averaging and tolerances

Modern scaled converters make use of large number of identical unit-capacitors or transistors instead of resistors. These are aggregated to form either hard-wired binary or linear scales. The cell-count grows exponentially with the number of bits, so the question arises as to whether the area is still workable or not. The answer is yes for the huge number of unit-elements required to implement the assistance of the high rank bits average out their errors. Consequently smaller unit-elements can be tolerated without compromising the overall accuracy, provided the cells' mismatch is equally distributed over the whole area occupied by the unit-elements.

Let us consider two 10-bit converters to illustrate the lesser impact of mismatch for higher rank bits: an R2R and a unit-element. A zero median Gaussian distribution with a standard deviation of 2^{-11} is assumed for the resistances of the R2R converter. Now let us consider the unit-element 10-bit converter. Since its MSB is implemented by 512 unit-elements in parallel, the standard deviation of this bit is equal to that of the unit-elements, called sigma, divided by the square root of 512. For the next bit sigma is divided by the square root of 256, and so on. Thus, every bit enjoys error averaging, but in different manners. The interesting point is that the most critical bit, the MSB, has the most error averaging while the less important, LSB, has none. If the two converters target the same standard deviation as far as their MSB, a sigma equal to $2^{-6.5}$ (1.1×10^{-2}) is enough for the unit-elements.

This is clearly illustrated in Fig. 2.12, which shows the *INL* and *DNL* curves of five binary weighted converters whose unit-elements have standard deviations equal to 0.5×10^{-2}, ten times the standard deviation of the resistors considered in Fig. 2.3. The *INL* is similar but appears more erratic, since errors increase as the rank of the bits decreases.

The cumulative curves shown in Fig. 2.13 correspond to five hundred 10-bit converters whose unit-elements exhibit standard deviations equal to 0.5, 1 and 2%. More converters comply with the one-half LSB specifications than in Fig. 2.4, notwithstanding the tenfold increase of the standard deviation because statistical averaging slightly overrules the impact of the unit-elements' relaxed tolerances.

It is generally assumed that the area of the unit-elements must vary like the reciprocal of the standard deviation squared. Thus, in the 10-bit converter

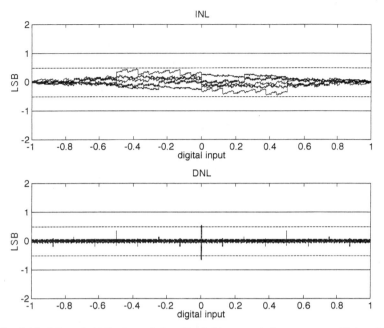

Fig. 2.12 *INL* and *DNL* plots of five 10-bit binary-coded converters utilizing 0.5% matched unit-elements. Notice the tenfold matching figure with respect to the example considered in Fig. 2.3.

above, the area of the unit-elements may be 512 times smaller than the area required by the same MSB element if it were to be implemented alone with the same accuracy.

So far we considered converters which use only binary weighted ensembles of unit-elements. As previously mentioned, it is possible to add also the unit-elements is such a way that they track the input data linearly, much like the height of the mercury column enfolds the temperature in a thermometer. Unlike binary-coded converters, ***thermometer coded converters*** require distinct accesses to every cell. Thus the circuit is more complex, defeating the idea that thermometer and binary-coded unit-element converters require the same area for the same resolution. A larger area is required, but the overhead remains manageable. Structured architectures, like those used in memories, can be used advantageously. The selection logic consists of a column decoder, row decoder and little logic in every cell. Thermometer-coded converters are still interesting devices as they offer superior *DNL* performances compared with binary-coded converters. Every step is controlled by a single unit-element so that the *DNL* ignores mid-code transition problems. The *DNL* is determined by the disparities among unit-elements and is thus a small fraction of one LSB

40 | Scaled D to A converters

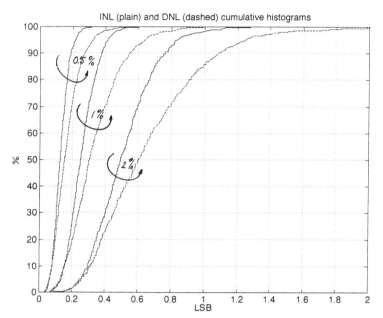

Fig. 2.13 Cumulative histograms (500 converters) of the max. *INL* and *DNL* of 10-bit binary-coded converters making use of unit-elements whose matching figures are equal to 0.5, 1 and 2%.

regardless of the resolution. However, this is not true for the *INL*. Mismatches affecting the unit-elements add up algebraically as the input evolves, so the transfer characteristic looks more erratic and is not symmetrical. Nonetheless, the *INL* is not worse than binary coded unit-elements converters, it is similar.

Figure 2.14 shows the *INL* and *DNL* curves of five thermometer-coded converters whose unit-element standard deviations are the same as those of the converters considered in Fig. 2.12. The most striking difference concerns the *DNL*, which is almost ideal. The erratic character of the *INL* is clearly visible but the magnitude is roughly the same as that of the binary-coded converters.

The max *INL* and *DNL* cumulative curves of three sets of 10-bit thermometer-coded converters, each five hundred samples wide, are displayed in Fig. 2.15. The unit-elements have standard deviations equal to 0.5, 1 and 2%. The *INL* plots are almost the same as those of Fig. 2.13 whereas the *DNL*, as expected, only reflects the tolerance of the unit-elements.

An interesting peculiarity of thermometer-coded converters is their smaller non-linear distortion. Because the number of unit-element aggregated per clock signal is strictly proportional to the difference between signal samples, the glitch energy varies like the signal slope (Chi-Huang and Bult 1998). Binary-scaled converters produce position dependent glitches, the mid-code

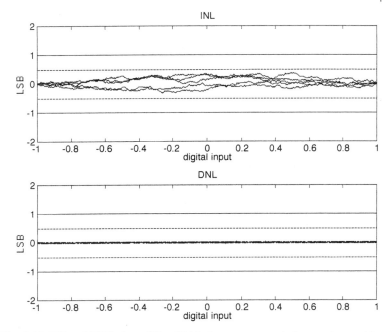

Fig. 2.14 *INL* and *DNL* plots of five 10-bit thermometer-coded converters making use of 0.5% matched unit-elements.

ones being the most severe. Thermometer-coded converters therefore produce less harmonic distortion, especially for slow varying input data.

Detailed consideration of two kinds of unit-elements converters, capacitive- and transistor-scaled, follow.

2.2.2 Capacitive scaling

Capacitors are currently implemented in MOS technology and are capable of performing many functions traditionally fulfilled by resistors. The quality of thermally grown oxide is exceptionally good; no other dielectric can compete. The gate of a non-volatile memory device, which is completely isolated from the external world by means of thermally grown oxide, can hold pico Coulombs for years as long as the oxide is not subject to U.V. radiation.

How are integrated capacitors fabricated?

Precision integrated capacitors generally consist of two highly doped polysilicon layers separated by a thin layer of oxide. The typical capacitance values per unit area vary from 0.5 to 1.5 fF/μm^2. Most capacitances rank between femto and a

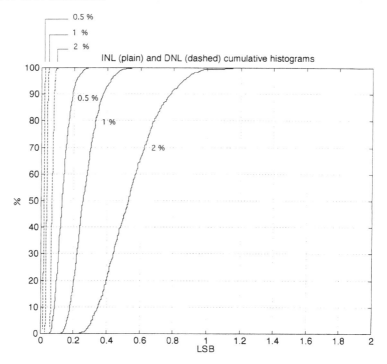

Fig. 2.15 Cumulative histograms (500 converters) of the max. *INL* and *DNL* of 10-bit thermometer-coded converters making use of unit-elements whose matching figures are equal to 0.5, 1 and 2%. Notice the small *DNL*.

few pico Farads. The lower plate is usually gate poly. The upper plate is deposited after thinning the oxide grown on the gate. This technology is actually an offspring of memory fab-lines after the first 16k RAMs were introduced. There are two other technologies: one where the top poly layer is replaced by an aluminum layer and another where the gate poly layer becomes the top plate and the lower plate is an implanted junction. In metal–poly capacitors, the oxide layer grown on top of the gate oxide is thicker than in poly–poly capacitors to avoid spiking which causes short-circuits. The potential capacitance per unit-area is therefore substantially smaller than with poly–poly capacitors. In poly–junction capacitances a special ion implantation through the gate oxide prior to the deposition of the gate poly is achieved in order to fabricate the junction. This is not possible in standard MOS technology, for the source and drain diffusions are fabricated after the gate poly has been deposited and etched away. Thus no junction is conceivable under the poly unless it is introduced before the poly deposition. Whichever technology is used, additional masks are necessary with respect to the standard MOS technology. The only capacitors that do not require additional fabrication steps are poly-inversion layer capacitors; that is, MOS transistors whose source and drain terminal are tied together to form one of the capacitor's terminals. The inversion layer plays the same role as the implanted junction above cited. Compared to poly–poly and metal–poly capacitors, the performance are poor as to parasitics, non-linearity, etc.

Parasitics

Regardless of the capacitor considered, parasitics are inevitably associated with the terminals. The bottom plate sees the substrate through either oxide or a depleted layer. The correspondent parasitic capacitance represents a non-negligible fraction of the nominal capacitance. The more efficient the isolation, or thinner the dielectric layer of the integrated capacitor, the smaller the parasitic with respect to the nominal capacitance. In practice, parasitic capacitances can be as large as 10 to 30% of the nominal capacitance. Nearly 60 to 80% are encountered with metal–poly capacitors for the relatively thicker oxide layer between metal and poly, which enhances the relative importance of the substrate. Poly–poly capacitors exhibit the lowest parasitic figures.

As to the top plate with respect to the substrate, parasitics are generally smaller since the bottom plate shields the top plate from the substrate. The remaining parasitics are due to fringing capacitances and the connections from the top plate to the rest of the circuit. The contribution of the latter is generally small since wiring is a small surface running over thick oxide. Thus the top plate parasitic capacitance with respect to the substrate is determined by the circuit layout, and is generally no more than a few fF.

Table 2.2 lists the matching performances of poly–poly and metal–poly capacitors as well as their temperature and voltage coefficients. Matching is a matter of size, as in resistors. Figures around 0.1 to 0.2 per cent are feasible for large devices. Smaller devices achieve 1% tolerances that are amenable for unit-capacitor converters. Ten-bit resolution converters are feasible but this is the upper limit, as in R2R converters. Linearity is another important consideration. Poly–poly and metal–poly capacitors offer the best linearity.

Table 2.2
Tolerances of various types of integrated capacitors (Gray et al. 1993)

Capacitor type	Absolute tolerance (%)	Matching tolerance (%)	Voltage coefficient (ppm/V)	Temperature coefficient (ppm/°C)
Poly–poly	± 10 to 30	± 1 area < 500 μm² ± 0.1 area > 10⁴ μm²	10 to 20	20 to 50
Poly–diff	± 10	same	20 to 100	50

44 | Scaled D to A converters

Highly doped poly layers behave like perfect conductors if they are in accumulation or strong inversion. However, even then a very thin partial depletion cannot be totally excluded which explains the small voltage coefficients listed in Table 2.2.

Capacitive divider pitfalls and challenges

Capacitive dividers have good and bad qualities, compared with resistive dividers. First their power consumption is drastically reduced as no DC current flows in the divider. The energy consumption only takes place during the charging and discharging cycles. Second, the DC and AC behaviors can be controlled independently. The output node of a capacitive divider may be precharged so that the output signal will be the attenuated replica of the input plus the pedestal fixed during precharge. This is not possible with resistive dividers. Third, capacitive dividers are strongly sensitive to stray-capacitance. All potential parasitic capacitances are highlighted in the divider shown in Fig. 2.16. The one at the input node in parallel with the signal source is unimportant since it only affects the input generator. The parasitic capacitance associated with the bottom plate of C2 has no effect since it is shorted. But the stray capacitance in parallel with the output node is of major significance to accuracy.

Basic capacitive converter architectures

Two capacitive D to A converters are considered in this section. The first is sensitive to stray capacitance, while the second is not. The first, shown in Fig. 2.17, is put to use in converters whose resolution is not to exceed 4 or 5 bits. The scaling network consists of an array of binary-weighted ensembles of unit-capacitors. The top plates are connected in parallel; lower plates are connected either to the reference voltage V_{ref} or the ground. The operation requires two cycles. First, the switch S shorts all the top plates to ground while the lower plates are grounded. Then, switch S is opened, leaving the top plates floating, at which time the lower switches take the states imposed

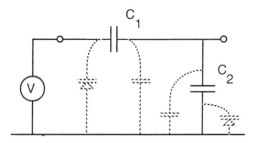

Fig. 2.16 The parasitic capacitance in parallel with the output impairs the performances of integrated capacitive dividers.

Unit-element scalers | 45

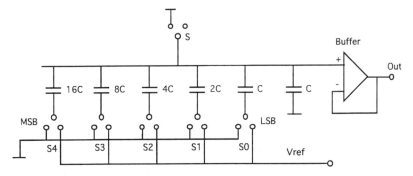

Fig. 2.17 The unavoidable stray capacitance at the voltage sensing node of this capacitive converter restricts its use to applications requiring no more than 4 to 5 bits.

by the input word. Each switch controls a divider whose ratio is fixed by the capacitance it enfolds over the sum of all other capacitances. For instance, if we consider the switch S4, its 16C capacitance in series, and the parallel combination of all other capacitances which amount to 16C, it is a divide-by-two network. When S4 is thrown from left to right, the buffer senses a voltage step equal to one-half the reference voltage V_{ref}. For S3, the series capacitance is 8C while the load reaches 24C. Consequently the output step equals $V_{ref}/4$. Since all the dividers share the same output node, the voltage sensed by the buffer represents the analog counterpart of the input code word.

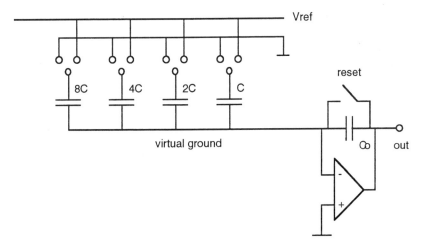

Fig. 2.18 A converter that is the capacitive counterpart of the resistive converter shown in Fig. 2.1.

Figure 2.18 shows the second converter, which, since it is stray-capacitance insensitive, is more suitable for resolutions reaching 10 bit. Here the lower plates are connected to the summing junction of an Op Amp while the other terminals are connected either to the ground or the voltage reference V_{ref}. The capacitor array and the feedback capacitor Co are discharged before the conversion. Once the reset is complete, the upper switches assume the positions dictated by the input code word. This produces a voltage step across Co whose magnitude is equal to V_{ref} multiplied by the ratio of the sum of all capacitances connected to V_{ref} over Co. The output represents thus the analog counterpart of the input coded word.

Stray capacitances do not affect this converter. The equivalent network of Fig. 2.19 shows that neither the capacitance associated with the input nor the output capacitor are of importance. Their charge and discharge currents ignore the summing node. The stray capacitance in parallel with the summing node of the Op Amp is not important either since it belongs to the artificial ground. Therefore the circuit of Fig. 2.19 is truly a stray-capacitance insensitive converter.

An interesting circuit is shown in Fig. 2.20, which combines both converters (Yee 1979). In section A, the voltage applied to the capacitors is constant while all the capacitances conform a binary scale like that in the converter of Fig. 2.18. In section B, the small horizontal capacitor connected to the summing node plays the same role as the capacitor connected to the ground at the end of the converter shown in Fig. 2.17. When the switch S5 connects the upper plate of the capacitor 8C to V_{ref}, V records a step equal to $V_{ref}/2$. The same reasoning applies for capacitor 4C, and so on. Every switch

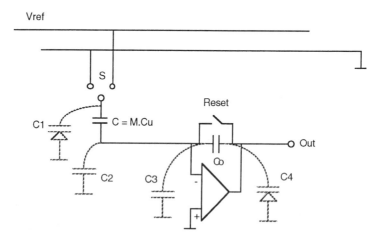

Fig. 2.19 The resolution of the converter shown in Fig. 2.18 can be much higher than that of the converter of Fig. 2.17 for it is insensitive to stray-capacitance.

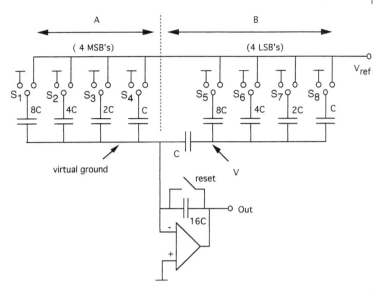

Fig. 2.20 The combination of the architectures illustrated in Fig. 2.18 and 2.19 can lead to substantial savings as far as silicon area.

in the B section defines a step, one-half the previous one. Thus, on the right the capacitance is constant, but the voltage is binary scaled. On the left, the voltage steps are constant but the capacitances change. One may argue that the converter of Fig. 2.20 is not entirely stray insensitive but this only applies to the less critical B section. The net result is a substantial reduction of the total capacitors area. The actual capacitors bank require only 12% of the requirements of a single 8-bit capacitive converter.

Layout aspects

For maximum minimization of the impact of mismatch sources, the layout of unit-capacitors must be carefully planned (Mc.Creary 1981, Shyu 1982). First, one may not ignore the fact that any unit-capacitor includes the short pieces of wires to adjacent capacitors. Though small, these must be taken into account. The way the capacitors are aggregated is also an important item. Figure 2.21 shows good and bad examples. The basic unit-capacitor with its adjacent interconnections is shown as 'a'. Layouts 'b', 'c', and 'd' aim at a ratio of four. Layout 'b' is the only correct one for it replicates four times the same unit-capacitor. In layout 'd' unnecessary bridges make the total capacitance larger than what is needed. Under 'c', although the area delineated by the dashed square equals four times the area of the unit-capacitor area, the capacitor 'c' is larger than expected once etching complete. This is due to the underetching, which concerns the periphery of the capacitors instead of the area. Another

48 | Scaled D to A converters

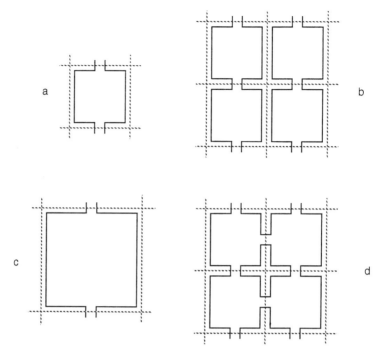

Fig. 2.21 Several layouts of integrated capacitors: the unit-capacitor (a), a correct fourfold capacitance (b) and two faulty layouts. In (c)the impact of under-etching is not taken into consideration and in (d) unnecessary bridges increase the capacitance.

potential source of mismatch is the difference in etching rates that results from inhomogeneous concentrations of the etching agents during the fabrication process. At the periphery of the bank of capacitors the concentration is less depleted than inside. To avoid this, one or two rows of dummies must surround the array of active capacitors so that the etching agents are depleted to the same extent throughout the critical capacitors. Finally a source of inaccuracy already mentioned is the existence of an eventual oxide thickness gradient over the wafer. Keeping the tolerances of a 30 nm thick oxide layer to within 0.1% implies that the variations of the layer thickness should be less than 0.03 nm. This is hard to achieve in practice, even over small distances. A technique that is currently advocated to counteract the effects of oxide gradients is the centroid distribution of capacitors. The idea is to compensate mismatches by combining unit-capacitors laid symmetrically around two perpendicular axes.

Since the layout is such a crucial item, capacitive converters make exclusive use of binary weighted architectures. The amount of parasitics that would be present with thermometer-coded converters makes this approach inpractical.

We will see soon that transistor scalers do not suffer from this drawback and therefore lend themselves equally well to both types of architectures.

2.2.3 Transistor scaling

Arrays of unit-current sources can play the same role as arrays of unit-capacitors. Each current source is a common emitter or a common source transistor controlled by the same base-to-emitter or gate-to-source voltage. The collector or drain currents are combined to implement the analog output. Both binary and thermometer coded converters are implemented this way.

An attempt to use bipolar transistor scaling arose earlier in this chapter when considering the resistive converter shown in Fig. 2.7. In the *LSB* section, the collector current scaling was done by sizing the emitters. The emitter–junction voltage controls the *collector current density*; thus, whenever the emitter area is multiplied by two, the collector current doubles. The same applies to MOS transistors as to their width over channel length ratio. This is not the best way to implement transistor scaling for emitters, or dissimilar *W/L*s, are sources of etching induced mismatches as in capacitor arrays. High-resolution transistor scalers therefore only use identical unit-transistors.

Evaluation of transistor mismatches

Transistor mismatches are the result of a number of items such as non-homogeneity of the etching process, edge effects, ion implantation, surface state charges, or mobility. Some of these are local effects, others relate to the distance between transistors. For instance, surface states cause local perturbations whereas oxide gradients modify the spatial mismatch distribution. The first introduce wide band noise; the second, variable bias. Extensive studies concerning the modelling of transistor mismatch are reported in Shyu (1984), Lakshmikumar (1986), and Pelgrom (1989). The last proposes a methodology that predicts technology induced mismatches affecting the threshold voltage V_T and the β factor ($\mu C'_{ox} W/L$). The corresponding standard deviations are expressed below as functions of the transistor active area and the distance between so-called 'matched' devices:

$$\sigma^2(V_T) = \frac{A_{VT}^2}{WL} + S_{VT}^2 \cdot D^2$$
$$\frac{\sigma^2(\beta)}{\beta^2} = \frac{A_\beta^2}{WL} + S_\beta^2 \cdot D^2 \quad (2.2)$$

The coefficients A_{VT}, A_β, S_{VT} and S_β are characteristic of the technology being used. Table 2.3 lists some of the coefficients reported in the paper by Pelgrom (1989).

Let us consider two examples of 'matched' n channel MOS transistors. These are two small and two large transistors, 5 µm wide 1.5 µm long, and 20 µm

Table 2.3
Coefficients used to predict the drain current mismatch in MOS circuits (Pelgrom et al. 1989)

Type	A_{VT} (mV / μm)	A_β (% / μm)	S_{VT} (μV / μm)	S_β (10^{-6} / μm)
n channel	20–30	1.5–2.5	4	2
p channel	25–35	3–4	4	2

wide 20 μm long, respectively. First, we ignore the distance separating them. We make use of a set of coefficients that are the averages of those listed in Table 2.3. For the small transistors, the threshold voltage standard deviation reaches 9.1 mV; for the large it is only 1.25 mV. The β standard deviations are respectively 0.7% and 0.1%. Next, we consider the same transistors separated by 500 μm. The additional contributions to the threshold voltage and β are respectively 2 mV and 0.1%. It is clear that the relative impact of the distance is much larger on the large matched transistors than on the smaller ones, which are overruled by local mismatches.

Evaluation of the mismatch impact on the unit-current sources shows that transistors driven by the same gate-to-source voltage exhibit a drain current standard deviation described by the expression:

$$\frac{\sigma^2(I_D)}{I_D^2} = 4 \cdot \frac{\sigma^2(V_T)}{GVO^2} + \frac{\sigma^2(\beta)}{\beta^2} \qquad (2.3)$$

GVO is the *gate voltage overdrive*, the difference between the actual gate voltage and threshold voltage. Accordingly to the above expression, mismatch is not only a function of the sizes and distances separating pairs of 'matched' transistors, but also of the gate voltage overdrive. The gate voltage controls the magnitude of the drain current according to a law that changes quadratic to exponential when going from strong to weak inversion. Therefore relatively large *GVO*s are required if we want to reduce the impact of mismatch. With a 0.2 V *GVO*, which corresponds to the onset of strong inversion, the small and large transistors above, separated by 500 μm, exhibit mismatches of respectively 9.3% and 2.4%. These drop to 6.3% and 1.6% when the *GVO* is raised to 0.3 V. Below 0.2 V, the mismatch increases too rapidly because the transistors are no longer in strong inversion.

D to A converter exploiting transistor scaling

The precision versus area tradeoff accounts for resolutions that do not exceed 10 bits. There are different methods to obtain higher resolutions, covered in the next chapter.

Unit-element scalers | 51

At present, both binary-scaled and thermometer-coded transistor-scaled converters are implemented. In the first, the drains or collectors of the unit-transistors are hard-wired to obtain sets of binary weighted currents. In the second, the outputs are added to track the magnitude of the input word. Larger input implies more transistors in parallel. This is what makes thermometer-coded converters monotonic and achieves good DNL performances regardless of the resolution. However, the need to access every unit-transistor entails more complexity. The cell selection requires row and column decoders such as RAM memories. A logic gate that acknowledges the coincidence of the signals delivered by the decoders is required in every cell. On the other hand, RAM organized converters are amenable to automatic design since they are regular structures.

The early thermometer-coded converter represented in Fig. 2.22 (Takahiro 1986) is an 8-bit CMOS device with a bandwidth of 80 MHz that was intended for HDTV applications. The converter scale consists of an array of 8 by 8 identical unit-transistor plus two smaller transistors, delivering respectively 50 and 25% fractions of the unit-transistors current. The row decoder operates along a thermometer-code while columns are selected alternatively left to right and right to left to circumvent oxide gradient effects to some extent. In more recent versions (Cremonesi 1989; Fournier and Senn 1990; Bastiaansen 1991; Kakamura 1991) the row and column decoders access every cell in pseudo-random sequences.

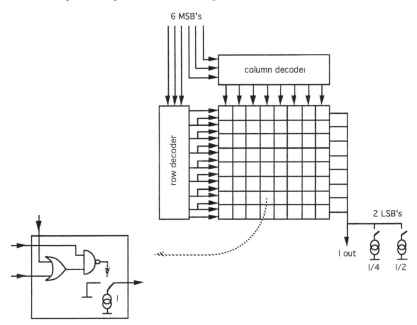

Fig. 2.22 An early version of an 8-bit 80 MHz D to A converter (Takahiro *et al.* 1986).

52 | Scaled D to A converters

Generally a cascode circuit is used to reduce the impact of Early effect on the output current and to minimize capacitive loading in the same time. Figure 2.23 shows the cascoded unit-transistor cell described in Chi-Huang and Bult (1998). The diode connected MOS transistor Q_1 drives the transistor Q_n, a member of the unit-transistors array. Both Q_1 and Q_n operate in strong inversion to avoid large mismatch figures typical of weak inversion. The drain of Q_n is cascoded via Q_c whose size and gate voltage are chosen in such a way as to maintain both transistors in saturation while minimizing the total voltage drop. The purpose of transistor Q_2 is to equalize the drain voltage of Q_1 and Q_n for matching. Note that the voltage drop across Q_2 is confined to a very small range, since the gate voltage of Q_1 determines the drain voltage of Q_2 whose source voltage may not desaturate Q_1. Taking a much larger W/L ratio for Q_2 than for Q_1 enables this. Q_2 consequently may not be in strong inversion but this is irrelevant since this transistor is current driven. The actual switch is a MOS differential pair that is tailored to maintain the voltage drop across its terminals within acceptable limits. Since the switch strengthens the cascode character of the circuit, the cell output impedance becomes so large that the output has no influence on the current. As a result, instead of an Op Amp, resistors can be used as output load. This substantially widens the converter bandwidth while it decreases the power budget. In the aforementioned reference a maximum sampling frequency of 500 MS/s is reported,

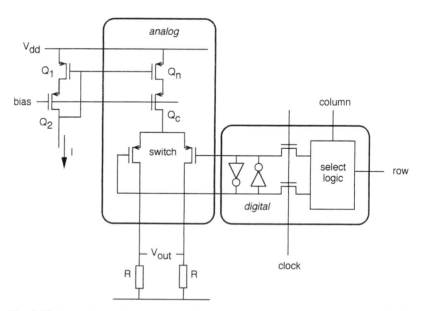

Fig. 2.23 Illustration of the analog-digital unit-cell section used in a modern 10-bit wide bandwidth thermometer-coded MOS converter.

which is good for telecommunications applications like digital sine-wave synthesis.

For optimal performance in resolution and area, combining both thermometer-coded and binary weighted architectures seems very promising: the *MSB*s control the thermometer- coded converter, and the *LSB*s the binary-coded. The area is smaller because fewer unit-cells must be accessed, but this is done at the expense of deteriorating the *DNL*. Since thermometer-coded converters have excellent *DNL* performance, there is enough room to withstand this. The real question is where to break the converter to obtain minimal total area while ensuring that both the *INL* and *DNL* comply with their specifications. This is addressed in Chi-Huang and Bult (1998) where the respective contributions of *DNL*, *INL* and extra logic to the total area are evaluated over a continuous scale that goes from full binary- to full thermometer-coded architectures. In binary-coded devices, the area is determined by the *DNL*, which is more or less twice as large as the *INL*. The *INL* tolerances are more relaxed than the *DNL*. In thermometer-coded devices, the area is controlled by the intra-cell logic. In between, there is a region where the *INL* dominates and overrules the growing logic contribution. In this region, one may take advantage of statistical error averaging. The optimal circuit is a combination where the *INL* and extra logic make similar contributions to the area. This defines the size of the largest and most economic thermometer-coded circuit, while it minimizes the impact of glitches. Generally the optimal converter consists of a low-resolution (two bit) binary-coded converter and a moderate-resolution (eight bit) thermometer-coded converter. In the 10-bit converter mentioned above, 256 current sources are arranged in a 16 by 16 RAM-structured array, and deliver each current equal to 4 LSBs while the binary-coded device covers lower contributions. Implemented in a 0.35 μm single-poly four metal layer technology, this converter only occupies 0.6 mm^2. It has an SFDR of 73 dB for an 8 MHz signal sampled at 100 MS/s, which drops to 60 dB for a 100 MHz sampled at 300 MS/s. The total power consumption is 125 mW for a 3.3 V supply.

High resolution parallel D to A converters

3.1 Introduction

Scaled D to A converters are not appropriate over 10 bits. Higher performances require matching conditions that are hardly attainable. In R2R converters, the tolerances surpass what is achievable and in unit-elements converters, despite the fact that the tolerances are more relaxed and the devices smaller, the element-count increases exponentially.

Two techniques are described in this chapter that enable enhancing the resolution. The first is known as the ***dynamic current division*** principle. It reduces the errors of the most significant binary weights by averaging out current mismatches over time. The second, specific to ***segment converters***, uses a single lower resolution D to A converter to encompass the conversion scale in several steps.

3.2 Current scaling using the dynamic current division principle

The *dynamic current division* principle improves current matching by averaging out differences between alleged equal currents over time. A factor of 10^3 to 10^4 improvement is claimed currently.

3.2.1 Dynamic current division principle

The dynamic current division principle is illustrated as the circuit shown in Fig. 3.1. It consists of two cascaded sub-circuits; the lower is a Widlar current mirror, the upper a bi-directional switch. Current is injected at the bottom and divided into two almost equal parts. To maintain their mismatch below 1%, two resistors are put in series with the emitters. In the upper circuit, the currents delivered by the Widlar circuit are exchanged periodically. If the time intervals, $T1$ and $T2$, were strictly identical, perfect matching of the output current would be achieved after low-pass filtering. In practice, a time mismatch of as little as 0.1% is feasible. Since mismatch of the output currents $I3$ and $I4$ is the product of the lower circuit mismatch multiplied by that of the time slots $T1$ and $T2$, an error as little as 0.5×10^{-4}, corresponding to 14 bits, is feasible.

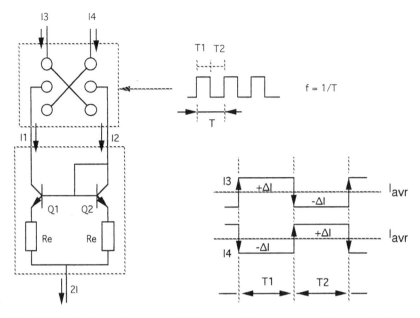

Fig. 3.1 In the dynamic current division principle, current matching results from averaging over time the outputs of a Widlar circuit.

A divide-by-two dynamic circuit is shown in Fig. 3.2. The current switches are cross-coupled Darlington pairs to reduce the base currents and lower the impact of their current gain mismatch.

3.2.2 High resolution D to A converter based on the dynamic current division principle

An early version of a converter using the dynamic division principle is shown in Fig. 3.3 (Van de Plassche 1976; Van de Plassche and Goedhart 1979). It consists of a number of cascaded replicas of the previous circuit. One of the two outputs of every cell completes the binary scale, the other is input to the next divider stage.

The reference current I_{ref} is mirrored in the first cell, consisting of two identical current sources. These are monitored by a negative feedback loop that compares the current from the reference I_{ref} to that of the right-hand current source and neutralizes the difference. Although the amplifier is a simple Darlington pair, the loop gain is large since the impedance at the input node of the Darlington pair is normally very large.

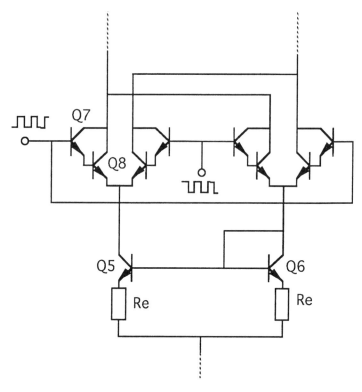

Fig. 3.2 Schematic of a bipolar divide-by-two dynamic current division cell.

Distinct clock signals must be applied to each divide-by-two cell. The smallest acceptable voltage across each one dictates the amount of clock signal level shifting between adjacent cells. With regards to the Widlar section, one V_{be} plus more or less 100 mV for the resistor is enough. For the Darlington pairs, a little less than $2V_{be}$s is necessary. Thus a total of about 2 V is required per cell. In the circuit shown in Fig. 3.3, the actual clock level shifting is done through blocks of three series diodes between each cell.

Since the number of cells and bits are the same, the total voltage drop across the entire level shifter may become a problem. A 16-bit converter made up of 6 divide-by-two cells and one 10-bit unit-transistors scaler requires 12 V. Because the voltage drop across a cell cannot be further reduced, the problem may be solved by increasing the number of outputs of each cell, as shown in the divide-by-four circuit in Fig. 3.4. This cell is an extended version of the previous Widlar circuit that divides the input current into four almost equal parts. These are cross-coupled in the upper cell, which runs under the control of four 'equal' time slots. Two of the four outputs are recombined to generate the halved input current; a third is the input current divided by four, and the

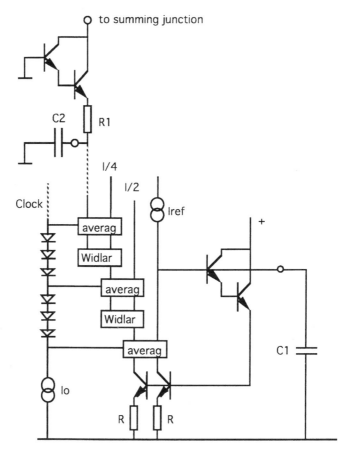

Fig. 3.3 In converters making use of the dynamic current division principle, cascading cells like the one shown in Fig. 3.1 obtain binary weighted currents.

last is the input of the next divided-by-four cell. The total voltage drop across the level shifter is now divided by almost two.

A converter using the latter circuit, described in (Schouwenaars *et al.* 1986), is shown in Fig. 3.5. The converter cascades three divide-by-four cells for the six *MSB*s, and uses a conventional unit-transistors scaler for the remaining 10 bits. The latter consists of 1024 Darlington pairs, each delivering a current of 62 nA. Unit-current sources are accessed accordingly to a pseudo-random pattern to counteract the impact of an eventual oxide gradient. The total voltage drop across the full converter ranges from 6 to 7 V. The reference current in this converter is a 4 mA bandgap current source with 200 ppm/°C temperature sensitivity. The binary scaled currents are filtered through external non-critical capacitors. The clock frequency is 200 kHz. Although the

58 | High resolution parallel D to A converters

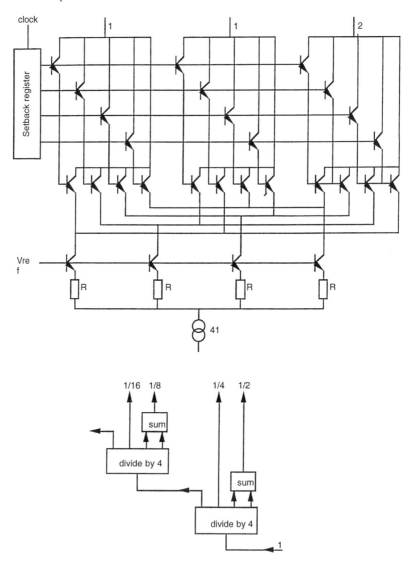

Fig. 3.4 The large number of cascaded cells required by converters that make use of divide-by-two cells calls for alternative circuits like this divide-by-four circuit.

converter was implemented using old standard bipolar technology, the area dedicated to the current weights required only 1.24×2.48 mm^2. The total chip area (including switches, read-out Op Amp and reference current source) is 3.5×5.5 mm^2. The 800 mW total power consumption is still very high compared with more recent MOS implementations discussed hereafter.

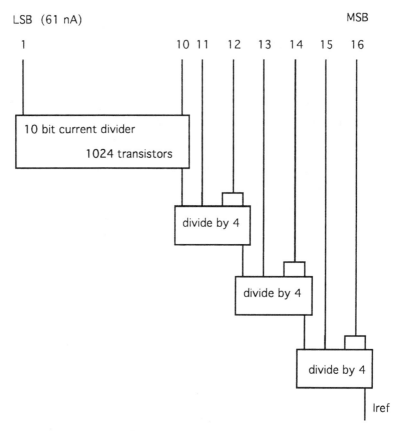

Fig. 3.5 The combination of bipolar divide-by-four dynamic current division circuits and a transistor scaler results in this 16-bit D to A converter that operates with a supply voltage of 6 to 7 V.

3.3 Segment converters

The name 'segment' refers to the fact that the converter transfer characteristic is the outcome of several concatenated lower resolution transfer characteristics. The same lower resolution D to A converter is used for each segment. The number of bits of the segment converter is the sum of the low-resolution converter number of bits plus the $\log_2$ of the number of segments. For example, a 16-bit segment converter may be implemented using a single 10-bit D to A converter and 64 segments. If segments are glued together perfectly so that no steps occur at the transitions, the *DNL* of this 16-bit converter is the same as that of the 10-bit internal converter.

Segment converters split the input code word into two fields: an *MSB* and an *LSB* field. The *MSB* word is nothing but a coarse quantized representation of the input, whereas the LSB word takes care of the fine quantization. The M-bit long *MSB* word controls the thermometer coded segmentation, while the LSB controls the actual binary coded converter. The output current or voltage is the sum of the fine and coarse converters. The thermometer-coded segmentation network generally consists of a set of calibrated voltage or current sources. When a ramp is applied to the converter these are connected to the internal binary converter one after another. Each time a new segment is activated, a new reference replaces the old one, the latter being added to those already addressed to keep track of the MSB word. Since every reference controls the gain of a segment, the overall transfer characteristic generally exhibits small erratic slope variations caused by mismatches of the segment references. This is clearly visible in the *INL* plot of the four-segment converter shown in Fig. 3.6, where small changes of slopes illustrate the segment

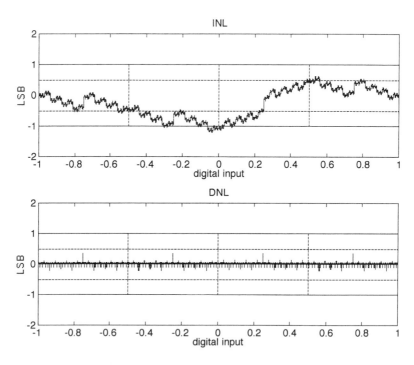

Fig. 3.6 *INL* and *DNL* plots of a 12-bit segment converter that consists of one 10-bit unit-elements converter and four segments. The standard deviations of the unit-elements and reference current sources are respectively 1 and 0.1%. The *DNL* is controlled by the impairments of the 10-bit converter whereas the *INL* combines the errors from the 10-bit converter and segment reference sources.

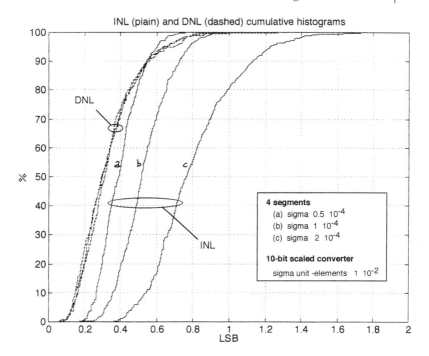

Fig. 3.7 Cumulative histograms of the max. *INL* and *DNL* of five hundred 14-bit converters, which consist of one 10-bit unit-elements converter and 16 segments. The unit-elements exhibit a standard deviation of 1% and the reference current sources respectively 0.005, 0.01 and 0.02%. The *DNL* plots are the same as expected whereas the *INL* plots depart from each other, illustrating the high degree of matching required as to the current references.

reference mismatches. Note that the impairments of the 10-bit internal converter are clearly replicated over the four segments. Unlike the *INL*, the *DNL* is not affected by the segment reference mismatches, since it is only controlled by impairments of the internal segment references.

The *INL* can be improved by implementing references with mismatch figures better than 10^{-3}. Three cumulative histograms are shown in the plot of Fig. 3.7, representing the maximum *INL* and *DNL* measured over a set of five hundred 14-bit segment converters, combining a 10-bit unit-transistors converter with 16 segments. The unit-transistors of the 10-bit converter have a Gaussian error distribution with 1% standard deviation while the reference sources, which define the 16 segments, have standard deviations of 0.005, 0.01 and 0.02%, respectively. The fact that internal converter impairments determine the *DNL* but not those of the references is shown through the accordance of the three *DNL* plots. This is not true for the *INL*. In fact, the *INL* imposes very strict tolerances. The plot shows that the acceptance rates based upon the plus or minus one-half *LSB* specifications are respectively 83, 45 and 7% for the segment standard deviations listed above. With regard to the *DNL*, the acceptance rates are all 86% notwithstanding the 1% tolerances of the unit-transistors. Of course, narrow tolerances such as those imposed by the *INL* are hard to achieve. But on the other hand, given the small numbers of segments in practice (64 is a maximum), the search for more sophisticated current sources with improved matching may seem justifiable. Section 3.3.4 shows that circuits like current copiers fill this gap very efficiently.

Note that the greater the number of segments, the greater the chance that *INL* errors compensate one another, provided mismatches affecting the references conform to a symmetric distribution around their average value. An oxide thickness gradient tends to defeat this statement of course but spatial randomization of the reference sources is a way to counteract this.

3.3.1 A typical current segment D to A converter

One of the earliest segment D to A converters is the 12-bit 'inherently monotonic' converter described in Schoeff (1979), shown in Fig. 3.8.

Although this converter performs poorly compared with more recent devices, its architecture clearly shows how transitional *DNL* degradation is being avoided in current segment converters. The internal D to A converter, controlled by the 9 lower rank bits of the input word, combines a 5-bit R2R scaler and a 4-bit transistor scaler. The remaining eight 'identical' current sources are the converter reference current sources. When the first segment is activated, the current source situated on the right feeds the converter while all other current sources are dumped to ground (Fig. 3.8 (middle)). The current delivered by the first source feeds the internal converter, which acts as a current divider. As more bits are activated, a larger portion of the current flows into the virtual ground of the Op Amp. Once the internal converter FS is attained, all the current from the first current source minus one *LSB* is fed to the summing node. Let us now assume that the input code word is increased by one *LSB* bit. As a result, the second current source is connected to the internal converter, which is reset to zero while the first current source feeds directly the summing node of the Op Amp (Fig. 3.8 lower part). Since the same current source feeds the summing node before and after switching, no transitional step is produced except for the incoming *LSB*. Repeating the same operation seven times, reaches the segment converter FS and all current sources connect to the Op Amp summing node.

Segment converters | 63

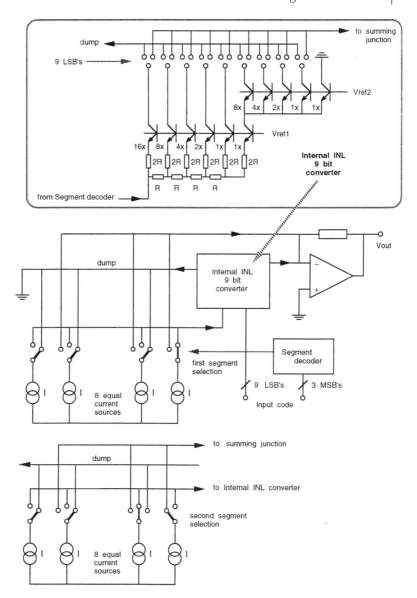

Fig. 3.8 In this early 12-bit segment converter, the 3 *MSBs* control 8 current reference sources and the 9 *LSBs* the actual converter.

3.3.2 A voltage segment D to A converter

The above converter is a current-mode segment converter. Voltage-mode segment converters are also feasible. They use equal resistors to implement voltage segment references, which supply the reduced resolution D to A converter. In Fig. 3.9 (Tuthill 1980), the divider consists of 16 resistors and the internal converter is a 12-bit laser trimmed R2R converter. The segment converter output is the sum of the voltage across the resistive divider plus the voltage delivered by the R2R converter.

The voltages applied to the internal converter terminals, Vx and Vy, are sensed by two unity-gain buffers to avoid loading the reference divider. Since the offset of the buffers are independent and easily reach a few mV, severe *DNL* could occur. Every time a segment is increased, the status of one of the divider terminals changes indeed from the high to the low terminal of the R2R

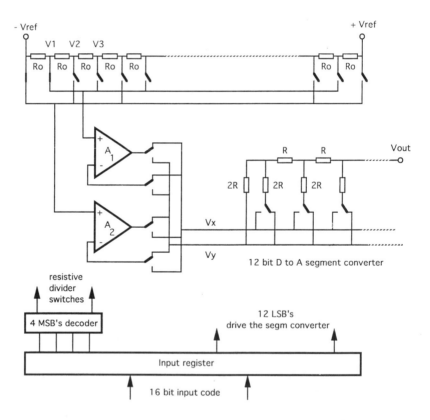

Fig. 3.9 Voltage segmentation instead of current segmentation is put to use in this 16-bit converter. Buffers are needed to avoid loading the resistive voltage divider that controls the voltage references.

converter. Consequently another buffer is connected in series with the same terminal. This problem is avoided if the buffers are interchanged every time segments are incremented. Thus, each node of the resistive divider 'sees' the same buffer amplifier and senses the same offset voltage. Since transitional steps are now ignored, the *DNL* is not affected by the buffers offset, but the *INL* is. The slopes of the individual segments reflect indeed the offset difference and produce a periodic zig-zag over the transfer characteristic.

Buffers can be avoided if the internal low-resolution converter is a capacitive, rather than an R2R, converter. No DC current flows out the resistive divider under steady state conditions. Thus the offset problem is avoided. A segment converter like that is shown in Fig. 3.10 (Fotouhi and Hodges 1979). This kind of architecture is extensively used in *Codecs* (see Chapter 4).

3.3.3 A 16 bit MOS segment D to A converter

The segment converter shown in Fig. 3.11 is a more recent device, described in (Schouwenaars *et al.* 1988). It is a current-mode 16-bit D to A converter, combining a 10-bit MOS transistor scaler and 64 'equal' current sources. It operates in the same way as the current sources segment converter above. Each current source has a tag defined by the segment decoder. Lower tag

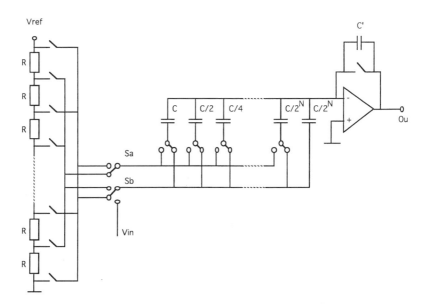

Fig. 3.10 The combination of resistive voltage segmentation with a capacitive A to D converter eliminates the need for buffers. The *INL* is improved in the same time because the offsets of the buffers of Fig. 3.9 are ignored.

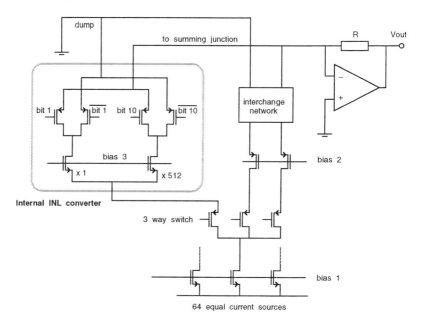

Fig. 3.11 In this 16-bit segment converter, which consists of 64 current references and one 10-bit converter, dynamic current matching is used to improve the *INL* performances.

current sources are connected to the summing junction of the Op Amp, and higher tag current sources are dumped. The one in the middle feeds the 10-bit converter.

Like the converters previously described, no spurious steps occur when segments are being concatenated. The particularity of this converter lies in its reference sources. Errors resulting from an oxide thickness gradient are compensated for by periodical exchange of the current sources, as in the dynamic current division algorithm considered at the start of this chapter. Figure 3.12 is a symbolic representation of the magnitudes of the segments current sources under the assumption that they exhibit a linear mismatch pattern. The situations depicted in 'a' or 'b' show two ways the current sources may be connected. Dark bars represent current sources already contributing to MSB words. The light bars designate current sources that are dumped, except for the one feeding the internal converter. If situations 'a' and 'b' are alternated for equal time intervals, the averaged currents disregard the gradient problem. This is a kind of spatial average that proceeds following the dynamic division principle described above. Hence, random access to the references is no longer necessary. However, the converter bandwidth is reduced since the output current must be low-pass filtered.

Segment converters | 67

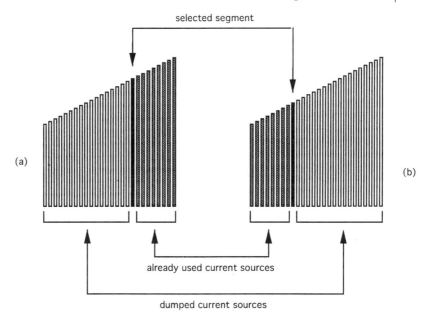

Fig. 3.12 Illustration of the strategy used in the converter of Fig. 3.11 in order to offset systematic errors caused by an eventual oxide gradient.

This 16-bit converter was implemented in 2 μm CMOS technology. Comparing it with the specifications of the segment converter previously considered, progress is clearly shown through its area and power consumption, respectively 5 mm^2 and 15 mW.

3.3.4 An improved version of the previous converter

The *INL* can be dramatically improved if the static current sources used previously are replaced by dynamic current copiers. An alleged 16-bit converter has been implemented this way with no need to periodically exchange current sources. This section explains the operation of the improved current sources.

Current copiers are MOS circuits that memorize analog currents (Daubert *et al.* 1988). The operation is shown in Fig. 3.13. During the sampling phase (left), the MOS transistor is connected like a diode with its gate and drain shorted. As current is injected, the voltage across the gate capacitance *C* grows until the drain and reference currents balance each other exactly. Once there is no longer any current flowing in the gate capacitance, steady state conditions are met. The second phase, the read-out phase (right), starts when the switch in series with the gate is opened. Since there is no path to discharge the gate

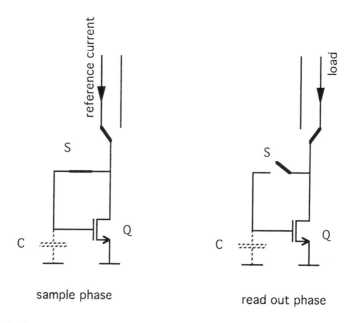

Fig. 3.13 During the sampling phase, current is being injected in the diode connected MOS transistor of the current copier. During the read-out phase, the gate is opened and the drain is connected to the load. The voltage stored across the input capacitance of the MOS transistor keeps the drain current unchanged.

capacitance, the gate-to-source voltage is 'frozen'. If we connect the drain terminal to another load, the drain current should remain the same. Thus from a user's point of view, current copiers memorize currents during the sampling phase and replicate them during the read-out phase. The operation is similar to that of current mirrors, but instead of two transistors, a single one serves as diode and output transistor accordingly.

This circuit is a very interesting precision analog circuit (Vittoz and Wegmann 1990) for the actual size and threshold voltage of the memorization transistor are irrelevant since the gate voltage adjusts itself to balance currents unavoidably. Thus if two current copiers are programmed with the same reference, their drain currents are identical regardless of threshold voltage and size mismatches. In other words, copiers behave like technologically independent current mirrors. However, a number of non-idealities impair their performance. The first impairment is leakage current which slowly discharges the gate capacitance. It is a consequence of the leaky junction that belongs to the series transistor switch. Generally its impact is small and may be ignored thanks to the periodic rewriting of the memorized data. The second problem is the finite output impedance of the memorization transistor, which should be large enough to make current insensitive to drain voltage changes. Long channels minimize this drawback, which is due to the Early effect, but a more efficient strategy is to cascode the output. The drain then senses only a very small fraction of the voltage modifications across the load. The third cause of non-ideality is the switch in series with the gate, which is another MOS transistor. When switched off, the gate-to-source overlap capacitance of this transistor induces an attenuated replica of the gate control signal on the memorization capacitance for the overlap capacitance and the memorization capacitor form a capacitive divider. Thus the voltage across the gate of the memory transistor experiences a small step, which slightly modifies the drain current. A similar additional perturbation is caused by the inversion layer charge of the MOS switch, which rather than disappear, divides itself between source and drain terminals. Part of this charge is dumped onto the memorization transistor and affects the drain current in the same manner as the overlap capacitance. There is no way to get rid of this but techniques to minimize the consequences exist and are reviewed hereafter.

Wouter *et al.* (1989) describe an improved version of the converter considered earlier that takes advantage of current copiers. He replaces the 64 reference current sources by an equal number of current-copiers. These are controlled by means of a single master current source, which is switched from one copier to the other. A 65th current copier is added to keep always 64 current sources available for the conversion and make the refreshing procedure transparent. Each of these 65 current copiers supplies 10 µA current. As a review of the circuit techniques believed to maintain their mismatch equal or lower than 5 nA, we consider the example described by Wouter *et al.* (1989). An N-type MOS transistor is used as the memory transistor. We consider a technology having a mobility µ of 600 cm²/V.s and a gate oxide capacitance per unit area C'_{ox} of 1.33 10^{-7} F/cm². With a gate voltage overdrive GVO or (VG – VT) of 0.5 V that keeps the memory transistor in strong inversion, the W/L ratio is derived from the expression of saturated drain current:

$$I_D = \mu C'_{ox} \frac{W}{L} \cdot \frac{GVO^2}{2} = 10 \text{ µA} \tag{3.1}$$

The result is a W/L of one. A large width W and a large length L are chosen to make the input capacitance of the order of 1 pF, to keep the voltage droop below 5 nA. This minimizes also the perturbation from injected charge of the series switch and keeps thermal noise at an acceptable level. Making W and L equal to 30 µm, yields a gate capacitance of:

$$C = C_{GS} = \frac{2}{3} C'_{ox} W L = 0.8 \text{ pF} \tag{3.2}$$

The saturation of the memory transistor is illustrated by the coefficient 2/3. The droop of the junction leakage can now be estimated, even though the size of the series transistor is still unknown but a reasonable leakage current of 1 pA is assumed. Starting from:

$$I_D - I_{Dref} = -g_m \frac{I_{leak}}{C} \cdot t \tag{3.3}$$

where the transconductance is given by:

$$g_m = \sqrt{2\beta I_D} = \sqrt{2\mu C'_{ox} I_D} = 40 \; \mu A/V \tag{3.4}$$

the maximum time allowed between refreshing can be found. Less than 100 µs is needed to keep the drain current droop below 5 nA. Refreshing at the stereo audio sampling rate of 44 kHz is thus sufficient.

Now let us estimate the perturbation resulting from the injected charge of the series switch. We assume the switching time is short enough to prevent the feedback loop around the current copier from controlling how the charge is split. Under these circumstances, 50% of the channel charge plus the overlap charge is injected into the memorization node. The charge balance thus sums up to:

$$(C_{ov} + 0.5 \; C'_{ox} \; W_{sw} L_{sw}) \; CK = \frac{2}{3} C'_{ox} \; WL \; \Delta V_{GS} \tag{3.5}$$

where CK represents the clock signal magnitude, C_{ov} the overlap capacitance, W_{sw} and L_{sw} the width and length of the series switch and ΔV_{GS} the perturbation at the memorization transistor gate voltage. The left-hand side expression represents the contribution of the series switch; the right-hand side that of the gate charge. The factor 0.5 supposes an equal split of the inversion layer charge from the series transistor, which is in the triode region when 'on'. The smallest possible dimensions should be chosen for the switch, e.g., 2 µm and 1.5 µm respectively for W_{sw} and L_{sw} since the final objective is to minimize the impact of the injected charge upon the drain current of the memory transistor. Smaller sizes are not advisable for they may compromise the speed. If we consider an overlap capacitance equal to 20% of the intrinsic gate capacitance, the gate voltage perturbation derived from eqn (3.5) is equal to 9 mV. This produces a drain current step of 360 nA using the transconductance magnitude predicted by eqn (3.4). However what is important is not the height of this current step but its mismatch since every current source is supposed to experience the same transient. We consider a 2% mismatch of the memorization transistor, which is a realistic assumption because it is a large transistor. For the series switch, which is a minimal device, a larger mismatch should be expected, nearly 10%. The global mismatch may reach thus 12%. Given the 360 nA current step, the drain current mismatch resumes to 43 nA, still 9 times too large!

To improve the performances of the current copier, the transconductance of the memorization transistor must be decreased. A smaller transconductance of the memorization transistor has another interesting consequence: it reduces the effect of the voltage droop across the gate capacitance. However since we do not want to lower the input capacitance nor the gate voltage overdrive, the drain current must decrease too. To keep the total current equal to 10 µA, another current source must be put in parallel with the memorization transistor. Figure 3.14 shows how this can be done. An auxiliary static current mirror is put in parallel with the copier. It supplies 90% of the total current. Thus the drain current of the memory transistor is divided by 10, and the W/L consequently divided by the same factor. To maintain the same input capacitance of the memory transistor, its width is divided by three and the channel length made three times larger. As a result, the transconductance is divided by almost ten and according to eqn (3.4) is equal to approximately 4 µA/V. With this tenfold decrease, the drain current step now reaches only 36 nA. However, the current copier mismatch worsens since its drain current is likely to change over much wider limits. The copier supplies the difference between the nominal 10 µA and the current delivered by the static current source. Matching of the latter is generally poor unless the current mirror is made very large. A 5% mismatch is a reasonable assumption. The current copier must then cancel out current differences as large as 50 to 150%. Hence, the transconductance of the memory transistor varies within plus and minus 30%. Added to the previous 10% from the series switch, the total mismatch of the output drain current now reaches 14 nA. It is possible to further reduce to

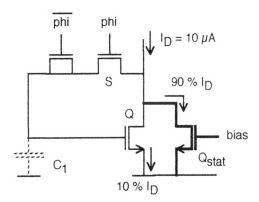

Fig. 3.14 The transconductance of current copiers must be very low in order to reduce the impact of charge injection. To compensate the concomitant drain current reduction, a static current source delivering 90% of the total current is put in parallel with the current copier. A dummy switch is put in series with transistor S moreover to further reduce the impact of charge injection.

some extend this threefold mismatch by making use of a dummy switch in series with the series switch. The dummy is a half-sized series replica of the MOS switch, the source and gate of which are shorted while the gate is controlled by the opposite clock signal. The idea is to cancel out more or less the equal but opposite charges of the series switch and the dummy. What remains determines the drain current mismatch, hopefully less than 5 nA!

The *INL* and *DNL* figures of the converter described by Wouter *et al.* (1989) are claimed to be 16-bit compatible, a spectacular result achieved with a 3 V supply and only 20 mW power consumption.

3.3.5 The current copier charge sharing problem

How the inversion layer charge is split between the MOS transistor terminals during cut-off is a problem that has been thoroughly studied in the literature. A brief survey of the most interesting results is presented in this section.

Let us consider the simple circuit of Fig. 3.15, which consists of a MOS transistor and two capacitors C1 and C2 (Wegmann *et al.* 1987). The voltages *V1* and *V2* across the capacitors are identical as long as the transistor is 'on'. When the transistor is switched 'off', the inversion layer charge Q_{inv} splits between the two capacitors. If these are identical, the charges are equally divided. If not, the trend observed at the start is the same, since the transistor is unaware of the difference between the capacitances. But since the voltage

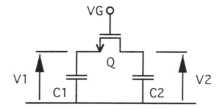

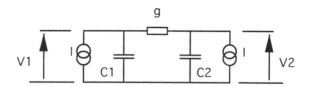

Fig. 3.15 While transistor Q is cut-off, its inversion charge is shared between the capacitors C1 and C2. The lumped circuit below models the mechanism controlling the actual charge division.

across the smallest of the two capacitor tends to deteriorate more rapidly than the other, current starts to flow in the transistor. This current tends to re-equilibriate the voltages. Since the voltage difference across the transistor remains small, the latter may be assimilated to a resistor with a conductance 'g' that decreases while the gate voltage drops. Hence, the charge re-equilibration mechanism progressively vanishes as the transistor nears cut-off. When cut-off is reached, the charge split freezes. The lower part of Fig. 3.15 shows an illustration of the lumped circuit that is supposed to give an account of the charge splitting mechanism. In the model, the two equal current sources in parallel with the capacitors illustrate the symmetrical charge split that takes place in the beginning. The magnitudes of these currents are easy to determine since they are equal to 0.5 C_{sw} (dV_G/dT) where C_{sw} represents the gate capacitance of the switch and dV_G/dT, the falling edge of the gate voltage and the re-equilibration mechanism is modeled by the variable conductance 'g', whose contribution is easily derived from the current linear approximation of drain current.

With this model, the actual charge split after the series switch has been cut-off may be predicted. The plot of Fig. 3.16 shows some of the conclusions presented in Wegmann *et al.* (1987). The variable B, plotted horizontally, is a measure of the switching driving force that relates the key parameters that

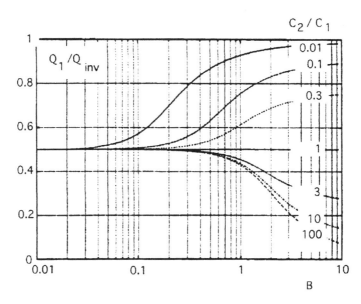

Fig. 3.16 After the cut off of transistor Q in Fig. 3.15, the charge left on C_1, can be determined from this plot. The variable B is defined in eq. (3.6).

74 | High resolution parallel D to A converters

control the charge splitting. The split is a function of the clock signal magnitude, the transistor size, the switching speed and ratio of the capacitances $C1$ and $C2$.

$$B = (V_{G\,\text{max}} - V_T)\sqrt{\frac{\beta}{a\,C_1}} \qquad (3.6)$$

$V_{G\text{on}} - V_T$: gate voltage overdrive before switching
β: beta of the transistor
a: rate at which the gate voltage falls (dV_G/dt)
and C_1: memory transistor gate capacitor

The amount of charge collected by C1 is displayed vertically, with regards to the total inversion charge Q_{inv} of the switching transistor.

When the gate voltage undergoes a sudden voltage drop, the variable 'a' becomes very large making B small; consequently, the re-equilibration process cannot take place and the inversion layer charge splits almost equally between the two capacitors. Consequently Q_1/Q_{inv} equals 0.5. On the other hand, when the switch operates at a slow pace, B is large and re-equilibration can take place. The voltages across the capacitors tend to be the same; thus the final charges are proportional to the capacitances.

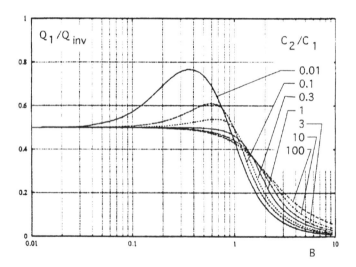

Fig. 3.17 The introduction of a MOS inverter in the circuit shown under Fig. 3.15 changes drastically the charge division mechanism. When the switching time becomes very long or B is large, the transistor introduces a negative feedback loop that tends to reduce the amount of charge left on the memory gate.

The problem is more complicated in the current copier because the voltage step across the input capacitance of the memory transistor is amplified, resulting in a large voltage swing across the series switch. The negative feedback loop formed by the memorization transistor and the series switch tends to maintain the voltage at the memorization node more constant to keep the drain and reference currents identical as long as possible. When the switch is nearing 'off' conditions, the loop-gain collapses and the re-equilibration mechanism ceases. The plot of Fig. 3.17 extends the results shown under Fig. 3.16 to the current copier (Macq and Jespers 1993). The meaning of variable B is the same, $C1$ represents the gate capacitance, and $C2$ is the drain parasitic capacitance. The memory transistor is modeled by means of an ideal transconductor whose g_m is equal to 220 µS. The most striking difference with respect to the previous graph occurs when B gets large or the cut-off mechanism is very slow. Since the drain and reference currents try to match each other as long as is possible through the feedback loop, the contribution of the inversion layer to the memory transistor gate becomes substantially smaller. Although this looks promising, its interest is marginal since the switching times required for large values of B are excessive.

Feedback A to D converters   4

Scaled D to A converters can be made A to D converters by interchanging their input and output terminals. This requires controlling the words driving the D to A converter so they balance the analog data fed to the A to D converter. The correspondent matching algorithm, the *successive approximation algorithm*, proceeds by cut and tries to progressively narrow the gap between the input signal and the analog counterpart of the output words. Since the algorithm proceeds bit by bit, one bit per clock cycle, from the *MSB* to the *LSB*, the entire conversion requires as many clock cycles as number of bits. Thus, successive approximation converters are not only slower than single clock scaled converters such as those considered in Chapters 2 and 3, they also run the risk that the input data drift more than one-half *LSB* during conversion. To avoid the loss of accuracy involved, the converters are forerun by a circuit that freezes the analog input data. This front-end is called a *sample-and-hold* (S.H.) circuit. This chapter starts with a few comments about this essential part.

4.1 The need for a sample and hold circuit

The sample and hold (S.H.) circuit fetches the input signal and stores it during full A to D conversion. Hence, input signal changes can no longer jeopardize the accuracy. No new sample may be fetched before the conversion is completed. Thus the converter fixes the *sampling rate*, not the S.H. circuit.

The generic S.H. circuit consists of a capacitor and a series switch, as shown in Fig. 4.1. As long as the switch is closed the voltage across the

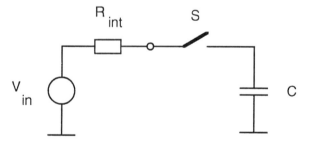

Fig. 4.1 A generic Sample-and-Hold circuit.

The need for a sample and hold circuit | 77

capacitor follows the input voltage V_{in}, provided of course that the bandwidth of the low-pass network formed by the source resistance R_{int} and the capacitor C encompasses the bandwidth of the input signal. When the switch opens, the voltage across the capacitor is frozen. It represents the magnitude of the input signal at the very moment the switch opened.

The meaning of 'opened' deserves some care. Switches are active devices, typically series MOS transistors. 'Opening a switch,' means that the impedance of the switch changes from a very low to a very large value in a very short time, typically an ns or less. While the switch is opening, the connection between the capacitor and the input weakens so that the final voltage across C is an average of the input signal throughout the opening sequence. To clarify the related *aperture time* concept, we examine the relationship between the time Δt and corresponding incremental magnitude of the sine wave represented in Fig. 4.2.

The most critical situation occurs near zero where the slope of the input signal is the steepest. This determines the smallest time Δt that corresponds to an incremental magnitude equal to one-fourth of an *LSB:*

$$\Delta t = \frac{2^{-(N+2)}}{\pi f} \quad (4.1)$$

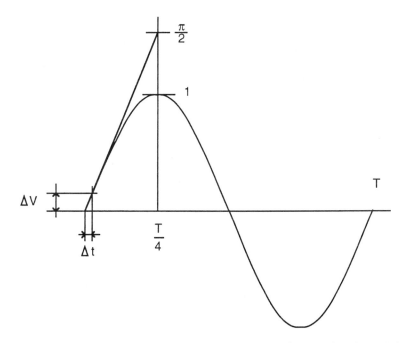

Fig. 4.2 The relation between the incremental magnitude ΔV and time interval Δt illustrates the concepts of aperture time and time jitter.

where N defines the converter resolution and f the frequency of the sine wave. An illustration of the above expression showing how Δt relates to the number of bits N and the frequency f is plotted in Fig. 4.3. Let us consider a few examples. First, let us assume that we sample a 20 kHz audio signal supposed to be quantized over 16 bits. The shortest time Δt equals 60 ps! Nearly the same result is found for a 5 MHz 8-bit coded video signal. A 10-bit converter, with a Δt of 1 ns implies a sine wave whose frequency does not exceed 78 kHz! Although these are somewhat pessimistic values since they don't account for the behavior of the real switch, it is clear that aperture times must be very short. If no, the exact instant of sampling suffers ambiguity.

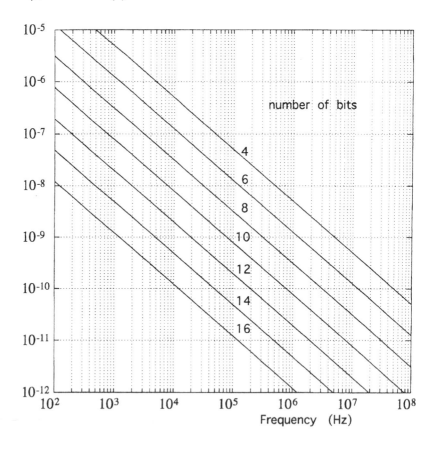

Fig. 4.3 Plot of eq. 4.1 that illustrates the relation between resolution, bandwidth and aperture time.

Sampling times are generally overruled by another impairment known as *jitter*, a generic term that covers erratic time modulations caused by noise and clock time-skew. Jitter is a very serious problem in high resolution high frequency converters. Its impact on the sampling time can be estimated using the same equation (4.1). If we consider a sine wave of magnitude A and assimilate jitter to a random variable with a standard deviation sigma, the noise power caused by jitter is given by:

$$\text{noise power} = \frac{A^2 \omega^2 \sigma^2}{2} \quad (4.2)$$

which determines the signal-to-noise ratio of the sampled data as:

$$\frac{S}{N} = 10 \log_{10}\left(\frac{1}{\omega^2 \sigma^2}\right) \quad (4.3)$$

Suppose we want to obtain a signal-to-noise ratio of 60 dB for a 10 MHz signal, which is equivalent of nearly 10 bits resolution. The standard deviation of the jitter determined using eq (4.3) is 16 ps.

Another important limitation of S.H. circuits using MOS transistors is charge injection, a problem previously cited in Section 3.3.5, about current copiers. When a MOS switch opens, charge is injected into the load capacitor from the evanescent inversion layer and clock signal feed-through. The resulting error is a small step that cannot be eliminated. Its magnitude is signal-dependent, which implies that it is likely to introduce non-linearity. Techniques that counteract the consequences of this induced offset are discussed in later sections.

Once the signal is caught, the S.H. circuit should maintain the magnitude of the sampled signal with the same degree of accuracy as the converter. However this performance is limited by a number of factors, such as leakage and feedthrough-noise. Leakage results from the junctions connected in parallel with the capacitor, but is not a major problem because of the actual sampling rates. Increasing the storage capacitor helps to slow the discharge, but sets severe demands on the switch series resistance. A larger capacitor requires a larger switch to keep the sample time, or *acquisition time*, constant. A larger switch concurrently increases parasitics, so a compromise must be made between speed and accuracy. Another problem is signal-feedthrough, not to be confused with switching noise. Signal feedthrough noise is due to the imperfect isolation between the input and output terminals of the open switch. Since the parasitic capacitance between its terminals is not zero, high frequency leakage is expected. The higher the frequency, the more difficult the isolation. Among the possible countermeasures, the most sensible is to short the signal ahead of the series switch when the latter stands open.

There are more elaborate sampling front-ends than the simple circuit of Fig. 4.1. Some take advantage of the fact that Op Amps are less sensitive to leakage and parasitics. They put the hold capacitor into a feedback loop around the Op Amp. Since the input is zeroed as a result of the feedback, leakage may be reduced substantially. Thus, a smaller hold capacitor is conceivable, making a smaller series switch acceptable. However, there is no change in switch parasitics since the injected charge and the hold capacitance are divided by the same factor. Feedback S.H. circuits are not appropriate as for converters, because the finite bandwidth of the Op Amp impairs the aperture time.

Most S.H. circuits are inherent parts of the converters they belong to; typical circuit aspects are covered in later chapters.

4.2 Sampled data spectral considerations

It is a well known fact that sampled data exhibit a spectrum forming a baseband, repeated at every multiple of the sampling frequency f_s. To reconstruct the original signal, repeated versions of the baseband spectrum may not overlap otherwise contributions from high frequency components fold back into the baseband and corrupt the sampled data. To maintain its integrity, the input signal must be band-limited to one-half the sampling frequency prior to sampling. This is done using a low-pass analog filter in front of the A to D converter called the *anti-alias* filter. Digital filters cannot be used for this task, as they themselves are sampled data systems. The low-pass filter exhibits constant gain until reaching its cut-off frequency, at which point the frequency response drops abruptly. The slope in the cut-off region fixes the order of the anti-alias filter. The steeper the slope, the higher the filter order. To determine the order of anti- alias filters, one needs to know two items: the cut-off frequency f_c and the attenuation DR_{dB} right at the half sampling frequency f_s. The order n of the anti-alias filter is given by:

$$n = \frac{DR_{dB}}{20\log_{10}\left(\frac{f_s}{f_c}\right)} \quad (4.4)$$

Accordingly to this expression, a 10th order analog filter is required if the frequency slope response must drop –60 dB in one octave. Although feasible, such a filter is very costly. If integrated, it may be a continuous gm-C filter, but this introduces non-linear distortion. Figures of over –70 dB, the equivalent of 12 bits, can hardly be achieved. Better performance requires external passive components. Chapter 7 shows that one of the most efficient ways to reduce the order of the anti-alias filter is to oversample the input signal.

Both Nyquist rated or oversampled signals undergo a frequency distortion inherent to the sampling procedure. The data are indeed stepwise approximations of the input signal, which is the same as filtering the input signal by a filter whose impulse response is rectangular. The magnitude equals the instantaneous value of the input signal, and the duration is the time T elapsed between consecutive samples. Such a filter has a frequency response whose absolute value is given as:

$$|H(j\omega)| = \frac{\sin\left(\frac{\pi f}{f_s}\right)}{\frac{\pi f}{f_s}} \qquad (4.5)$$

The resulting frequency distortion consists of a main lobe whose magnitude progressively deteriorates from one to zero near the sampling frequency f_s, plus a number of smaller lobes located between every multiple of the sampling frequency. According to the expression above, the attenuation reaches 3.92 dB at half the sampling frequency. This distortion may be compensated by introducing inverse pre-distortion within the anti-alias filter.

4.3 A to D converters based on feedback loops

Feedback converters were the foremost architectures advocated for A to D converters because they capitalized readily available expertise with D to A converters. As mentioned previously, the feedback loop controls the analog output of a slave D to A converter to match the unknown analog input. The architecture is shown in Fig. 4.4. The resolution depends not only on the accuracy of the internal D to A converter but also on the loop gain and feedback loop offset. If the loop gain and offset are respectively infinite and zero, the performances of the A to D converter are the same as those of the slave A to D converter, interchanging the roles of input and output ports. The *INL* and *DNL* characteristics are identical too, due to the distinct definitions given in Chapter 1 for both types of converters. The feedback loop enfolds a comparator followed by a dedicated logic block, which determines the words controlling the D to A converter. An appropriate algorithm runs the logic block that interprets the data from the comparator and assembles the words driving the D to A converter bit by bit, from the *MSB* to the *LSB*, one bit per clock cycle. When the *LSB* is set, the code word driving the D to A converter is declared the 'digital counterpart' of the analog input.

The logic that implements the algorithm controlling the words driving the D to A converter generally mimics the weighting of ponderous material.

82 | Feedback A to D converters

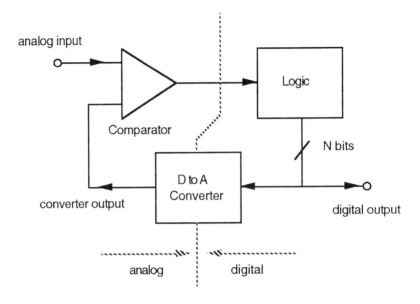

Fig. 4.4 The feedback loop around a scaled D to A converter changes the latter in successive approximation A to D converter.

Unknown weights are first checked against the *MSB* or 0.5 kg reference. Then, depending on the outcome, they are checked against either (0.5 + 0.25) kg or 0.25 kg, and so on. This procedure is repeated until the smallest weight, the *LSB*, is reached. Even though ponderous weights are not binary-scaled in practice, the similarity is obvious.

The plot of Fig. 4.5 illustrates the weighting algorithm unfolding a 0.65 V input voltage with a 1 V full scale reference. First, the input signal is checked against half the reference $V_{ref}/2$ or 0.5 V. Since the input is larger than the reference, the next weight $V_{ref}/4$ is added to $V_{ref}/2$. The D to A output then equals 0.75 V, which is too much. Thus, the $V_{ref}/4$ reference is erased and the unknown tested against $V_{ref}/2 + V_{ref}/8$ or 0.625 V, and so on. After 10 steps, the output voltage of the D to A converter reproduces the unknown analog input within plus or minus $V_{ref}/2^{10}$. The D to A output is the 10-bit fractional representation of the input voltage divided by the reference V_{ref}.

As in any other feedback loop, convergence implies that the loop gain be negative. Since non-monotonicity may be interpreted as a local change of the sign of the transfer characteristic, this kind of impairment should be avoided, otherwise the logic controlling the converter will endlessly repeat the same sequences without ever reaching the code word that represents the input.

A to D converters based on feedback loops | 83

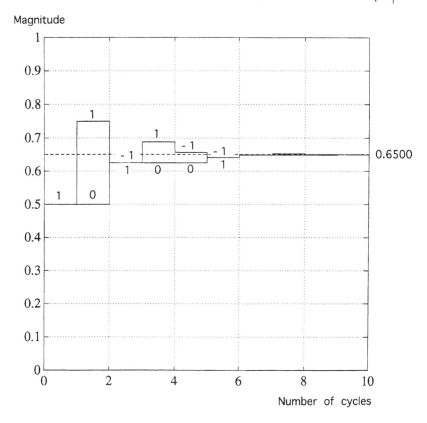

Fig. 4.5 Illustration of the weighting algorithm.

4.3.1 The charge redistribution A to D converter

As noted at the start of this chapter, the vast majority of feedback converters are straightforward implementations using the D to A converters considered in Chapters 2 and 3. These were described in papers published one or two years after the corresponding D to A converters were introduced. This section only covers converters whose architectures have a distinct feature worthy of mention. Charge redistribution converters (Creary and Gray 1975) are a good example.

Charge redistribution converters are basically feedback converters using the stray capacitance sensitive D to A converter considered in Fig. 2.17. The converter shown in Fig. 4.6 has three successive modes of operation. First, the analog input signal is sampled. This grounds all the upper plates of the capacitors identified as the switching level of the comparator. At the same time, the lower plates are connected to the input signal, with the exception of

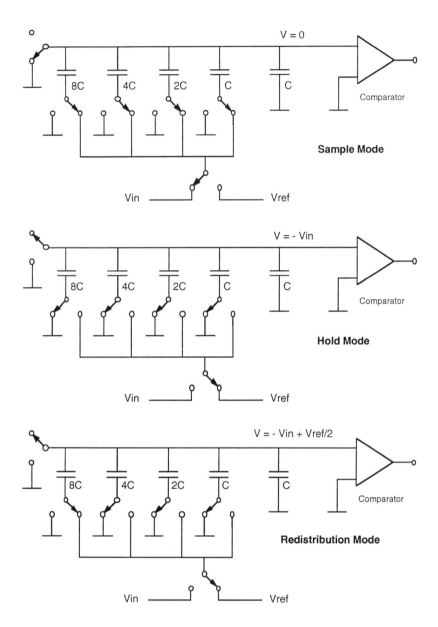

Fig. 4.6 Three phases are required to run charge redistribution A to D converters.

the extreme right-hand capacitor that remains grounded. After this, the 'hold' mode is entered. All upper plates are left floating while the lower plates are grounded. The voltage V of the upper plates jumps from zero to $-V_{in}$. This action ends the sample-and-hold procedure. The converter then enters the 'redistribution' mode. First the lower plate of the *MSB* capacitor (the 8C capacitor in the example) is connected to the reference voltage, so that a step equal to $+V_{ref}/2$ is superposed onto the upper plates. The comparator senses the voltage which is now $(-V_{in} + V_{ref}/2)$. Depending on the result, the logic block determines whether the switch controlling the *MSB* capacitor should be grounded or maintained in the same position. Similar decisions are made for each following capacitor until the *LSB* switch is set. When the last capacitor is reached, the voltage across the upper plates is smaller than one *LSB*. This is an important point, for the parasitic capacitance that impaired the accuracy of the D to A converter (see Section 2.2.2) is now fully discharged and its influence thus fully neutralized. Consequently, the converter is not affected by the parasitic capacitance of the upper node (contrary to the D to A converter).

When the conversion process ends, the charge initially distributed among the capacitors (except the most right-hand one) has been redistributed among only those capacitors whose lower plates are connected to V_{ref}. The others are discharged. This feature led to the name '*charge redistribution*' given to the converter.

4.3.2 A simple comparator

Comparators are moderate gain wide bandwidth amplifiers with low offset. Unlike Op Amps, they aim for speed and ignore stability. In Op Amps, the bandwidth is traded off for phase margin to yield stability. In comparators, phase margin and pole-zero constraints are ignored. Thus comparators are distinct, even though they share a common objective with Op Amps: the minimization of an error signal in a feedback loop.

Let us consider the circuit shown in Fig. 4.7, which represents the dynamic comparator described in Creary *et al.* (1975). The comparator is made up of three stages: two cascaded inverters, a differential amplifier, and a latch that delivers the binary output. A MOS transistor shorts the input and output of the first inverter A1. When 'on', the input and output take the same voltage V_o, located somewhere near the middle of the transfer characteristic shown in the plot of Fig. 4.8. Since the second inverter A2 is identical to the first, its output voltage will be the same. Both inverters are in a region where their gain is the largest, typically 30 to 50 dB. As long as S is closed, the input node of A1 exhibits very low impedance, for the short around this stage turns the input node into a kind of virtual ground. Its impedance may thus be of the order of a few hundred Ohms. As a result, the inverter and the switch S play the same role as the upper switch in the charge redistribution converter shown in

86 | Feedback A to D converters

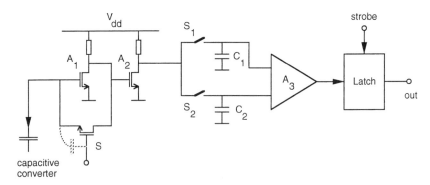

Fig. 4.7 Schematic representation of an early version of a MOS comparator intended for successive approximation A to D converters.

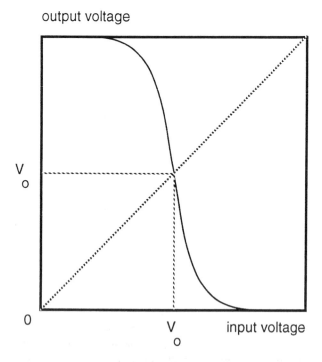

Fig. 4.8 The transfer characteristic of the two first stages A_1 and A_2 in Fig. 4.7.

Fig. 4.6. Instead of being shorted, the voltage of the upper node is set at V_o, which is the reference against which the comparator operates. Now the converter may enter the 'hold' and 'redistribution' modes. The switch S opens. The input impedance of the comparator changes from a small to a very large value. The two inverters behave as a single stage. The voltage input to the comparator now undergoes the sequence considered above, from V_o to $(V_o - V_{in})$ and from there to $(V_o - V_{in} + V_{ref}/2)$. Since the result is checked against V_o, the signal at the output of A2 is the magnified representation of the outcome of the first test.

The differential stage following the inverters A1 and A2 is intended to compensate the residual offset. When the feedback switch opens, the input node of the comparator undergoes a small voltage step due to charge being injected from the switch S. Although quite small, this charge produces a step that is amplified by the first and second inverters and is superposed on the data. The task of the differential stage A3 is to compensate for this error. Therefore the signal delivered by A2 is stored across the capacitor C1 after the switch S opens but before the converter enters the 'hold' mode. Once this is done, the 'hold' phase starts. The output of the inverter stage is now stored onto C_2 via the switch S_2. Since the same offset is stored across C_1 and C_2, the perturbation becomes a common mode signal, which will be ignored by the differential amplifier. The only offset remaining is that arising from the charge injection mismatch of the switches S_1 and S_2. Referred to the input, it is divided by the gain of the first stage. Hence, the offset may be as small as, or less than, 1 mV.

4.3.3 Auto-correction

The relative simplicity of charge redistribution A to D converters makes them attractive, in spite of their large number of unit-capacitors. Accuracy of up to 9 or 10 bits can be achieved without too much difficulty. As usual, the unit-capacitor mismatch sets the performance unless auto-calibration is applied. This section covers a typical auto-calibration algorithm well suited for these converters that is similar to the procedure used in the D to A converter described in Section 2.1.5.

The error evaluation algorithm of the 15-bit converter described by Lee *et al.* (1984) is shown in Fig. 4.9. First, the largest capacitor is checked by sampling the reference voltage V_{ref} on the lower plates of all the capacitors, with the exception of the *MSB*, which is discharged. This is shown in the upper part of Fig. 4.9. In the lower part of the same figure all the switches have been reversed once the upper plates were freed. The *MSB* and the other capacitors in parallel now form a divide-by-two network. If the *MSB* capacitor were equal to the sum of all the other capacitors, the voltage of the upper

88 | Feedback A to D converters

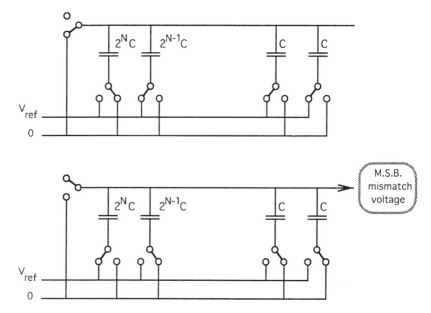

Fig. 4.9 In the auto-calibration algorithm reported in (Lee *et al.* 1984), every capacitor is checked separately. The pre-charge phase above is followed by the evaluation phase below.

nodes would be zero according to the previous precharge mechanism. Thus any departure produces a residual voltage that is a direct measure of the mismatch. The same test is repeated with the next capacitor, excluding the *MSB* capacitor. When the *LSB* capacitor is reached, the collected data are used to evaluate the errors of the binary array. It is then possible to determine corrections for every bit.

The circuit shown in Fig. 4.10 represents an auto-calibrated successive approximation 15-bit converter that uses the correction algorithm above. The converter consists of a 10-bit charge redistribution D to A main converter supplemented by a fine 5-bit thermometer-coded sub-converter. The latter is a divider network formed by 32 resistors whose output voltage is applied to the bottom plate of the second unit-capacitor counted from the right in Fig. 4.10. The 15-bit converter is similar to the segment converter shown in Fig. 3.10, but with the capacitive and resistive parts interchanged. During the auto-calibration phase, the 10 *MSB*s of the charge redistribution converter are checked using the procedure described above. Errors are evaluated indirectly by an auxiliary 7-bit converter, which is assumed to zero every output of the main converter. Thus the data controlling the auxiliary converter are digital counterparts of the errors of the charge redistribution converter. Once the

A to D converters based on feedback loops | 89

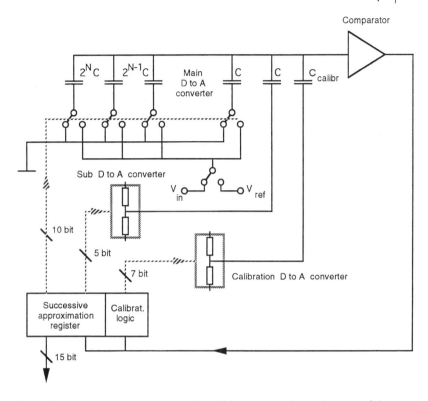

Fig. 4.10 Schematic representation of a 15-bit converter that makes use of the auto-calibration algorithm shown in Fig. 4.9.

calibration is completed, the correction codes to apply to the 5-bit converter are computed and stored in an on-board memory.

A more sophisticated calibration procedure is contemplated for high resolution converters, to cancel out errors from thermal noise sources inside the converter. Noise easily interferes with the calibration procedure. A 15-bit converter with a 5 V dynamic range has a step size of only 153 µV. A way to reduce the impact of noise is to repeat every test several times and only derive correction data from averaged readings. For instance, in the 15-bit converter considered above, where the reported noise level reaches 40 µV r.m.s, every test is repeated 16 times. Of course, this raises the question: what is the value of an alleged 15-bit resolution under these circumstances? The answer depends on whether static or dynamic performances are considered. In dynamic performance, the thermal noise is averaged out over time so that its impact is a slight increase of the noise floor, which affects the *ENOB*.

4.4 Codecs

Not all converters, whether D to A or A to D, are strictly linear devices. In some applications, non-linear transformations are performed to enhance the efficiency of transmission media. Pre-distortion of the input signal prior to digitization is followed by post-distortion to resume the linearity of the transmitted signal. *Codecs* are such devices, currently used in telephony. They are intended to digitize voice signals into 8-bit coded words, which correspond to a signal-to-noise ratio of 50 dB. Eight bits guarantee intelligibility for full scale signals. However voice signals may change by almost 30 dB from one speaker to another, even for the same speaker. Since the signal from the weakest speaker must be coded into 8 bits, the 30 dB overhead required for the most noisy speaker implies that 13 bits should be considered: a substantial waste of bandwidth.

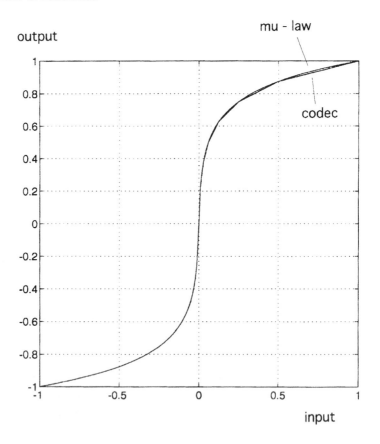

Fig. 4.11 Ideal (continuous line) and 8-bit approximate (broken line) CODEC characteristics.

Two international pre- and post-distortion laws have been issued, which yield 8-bit code words for compressed telephony voice signals. One is the μ-255 law as defined by CCITT recommendations [CCITT 92], summarized in the following (x and y are the input and output, respectively):

$$y = \frac{\ln(1+\mu|x|)}{\ln(1+\mu)} \qquad (4.6)$$

In practice, the pre- and post-distortions are approximated by means of a finite number of concatenated linear segments whose gain (or slope) is divided by two every time the magnitude of the signal jumps from a segment to the next. Two sets of eight segments are implemented, one for positive and another for negative signals. The resulting transfer characteristic is a broken line whose breakpoints lie on the ideal μ-law as shown in Fig. 4.11. Each segment is divided in 16 equal steps to yield the total of 8 bits. If the pre- and

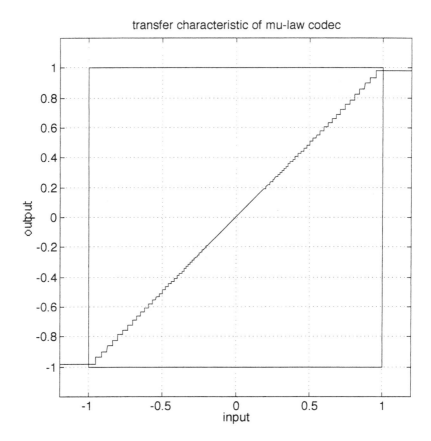

Fig. 4.12 When pre- and post-distortions are combined, a linear transfer characteristic with variable step size is obtained.

92 | Feedback A to D converters

post- distortion are combined, the transfer characteristic looks like a linear transformation with a step size that varies almost continuously over the full scale as shown in Fig. 4.12. As a result, weak and large signals are digitized with almost the same resolution. The signal-to-noise plot shown in Fig. 4.13 confirms this. Signals whose magnitude ranges from 0 to –30 dB show more or less constant 50 dB *SNR*. The polygonal curve shown in the same figure illustrates CCITT recommendation G.712 (09/92) proposed for pulse code modulators. Since the *SNR* encompasses the CCITT specification, intelligibility is achieved.

Segment converters are naturally gifted for the implementation of Codecs. D to A conversion is done by a segment converter like that shown in Fig. 3.10. The inverse transformation is the result of a feedback loop around the segment converter. Ohri *et al.* (1979) present an interesting approach for the D to A conversion. A digital translator transforms the 8- bit input coded words into 13-bit words that drive a binary-coded capacitive converter. The latter exploits the architecture shown in Fig. 2.20, which reduces the number of unit-capacitors through the use of a 7-bit *MSB* field and a 6-bit *LSB* field. Of course

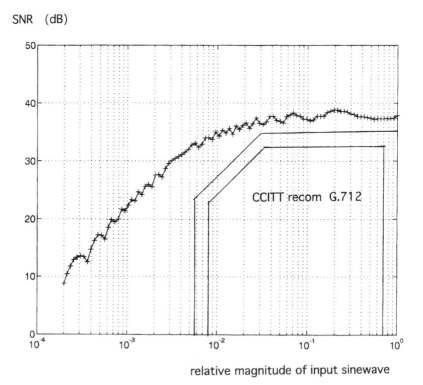

Fig. 4.13 Signal-to-noise plot versus magnitude of a sine wave after undergoing the combined pre- and post-distortion illustrated by Figs 4.11 and 12.

neither the *INL* nor the *DNL* comply with the 13-bit resolution. However, this is not important since the codec resolution must not exceed 8 bits. The interesting point lies in the digital decoder that substitutes 13-bit words for 8-bit incoming words. No post-distortion is required for the 4 lower rank bits since these are true binary representations by themselves. For the 4 MSBs, the problem is different. The highest rank bit determines the sign of the input and the three remaining bits yield the segment number.

Algorithmic A to D converters

5

Algorithmic converters, also called cyclic converters (McCharles *et al.* 1977), are straightforward implementations of the binary division algorithm. They output the binary fractional expansion of the input divided by the constant that fixes the conversion dynamic range. One of their key features is their potential for low-power consumption (Chen and Wu 1998). Currently the resolution ranges from 13 to 14 bits. The conversion speed is the same as that of successive approximation converters since identical numbers of clock cycles are required to yield the same resolution. A new bit is generated each time a new cycle is launched, from the *MSB* until the *LSB*. However the algorithm proceeds differently: for each clock signal the input data are updated; in successive approximation converters the input remains unchanged and the reference weights are adjusted. Since the same operation is repeated invariably cycle after cycle, a single common block of hardware is required regardless of the resolution. Above 10 bits however, the size increases for matching and compensation techniques become essential.

Algorithmic converters belong to a more general class of multi-step converters, which are covered in Chapter 8, under the sections dealing with multi-bit recycling and pipelined converters.

5.1 The cyclic algorithm

Figure 5.1 illustrates the flow-graph of the cyclic algorithm.

All the R_i's are *remainders* or *residues* from the previous cycles. The first residue R_1 is the input signal itself. Every new residue is derived from the former according to the relation:

$$R_{i+1} = R_i - b_i \cdot V_{\text{ref}} \tag{5.1}$$

where b_i is equal to + or −1, determined by the sign of the previous residue.

Table 5.1 displays the conversion steps of the number 0.170 considering a reference of 1. The second and third columns represent the residues and the bits b_1 respectively generated after each cycle. Since the first residue (the signal) is positive, the first bit b_1 is +1. The second residue R_2 is −0.66, the next −0.32, and so on. The sequence listed in the last column displays the reconstructed output data after multiplying the bits in the third column by

The cyclic algorithm | 95

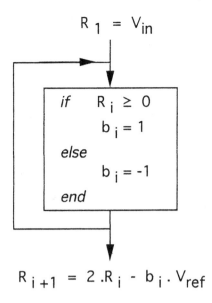

Fig. 5.1 Flow graph of the non-restoring algorithm used in cyclic converters.

Table 5.1
Algorithmic conversion of the number 0.170

order i	R_i	b_i	cumul. Σ
1	0.1700	1	0.5000
2	−0.6600	−1	0.2500
3	−0.3200	−1	0.1250
4	0.3600	1	0.1875
5	−0.2800	−1	0.1562
6	0.4400	1	0.1719
7	−0.1200	−1	0.1641
8	0.7600	1	0.1680
9	0.5200	1	0.1699
10	0.0400	1	0.1709
11	−0.9200	−1	0.1704
12	−0.8400	−1	0.1702
13	−0.6800	−1	0.1700
14	−0.3600	−1	0.1700
15	0.2800	1	0.1700
16	−0.4400	−1	0.1700

their respective fractional weights $1/2^i$ and taking the cumulative sum. Column 4 clearly reproduces R_1 as the cycle count continues. After the tenth cycle, the difference is less than 0.5% and after the 13th cycle, less than 10^{-4}.

The algorithm is easy to understand through the following:

$$R_1 = V_{\text{in}}$$
$$R_2 = 2R_1 - b_1 \cdot V_{\text{ref}}$$
$$R_3 = 2(2R_1 - b_1 \cdot V_{\text{ref}}) - b_2 V_{\text{ref}} \tag{5.2}$$
$$R_4 = 2\{2(2R_1 - b_1 \cdot V_{\text{ref}}) - b_2 V_{\text{ref}}\} - b_3 V_{\text{ref}}$$

or

$$R_4 = 2^3 \{R_1 - (b_1 \cdot 2^{-1} + b_2 \cdot 2^{-2} + b_3 \cdot 2^{-3}) \cdot V_{\text{ref}}\}$$

thus:

$$R_{i+1} = 2^i \cdot \left\{ R_1 - \sum_{j=1}^{i}(b_j \cdot 2^{-j}) \cdot V_{\text{ref}} \right\} \tag{5.3}$$

and for a large number of number of cycles:

$$\lim_{i \to \infty} \left\{ \sum_{j=1}^{i}(b_j \cdot 2^{-i}) \right\} = \frac{R_1}{V_{\text{ref}}} = \frac{V_{\text{in}}}{V_{\text{ref}}} \tag{5.4}$$

5.1.1 The Robertson plot

The Robertson plot is a graphic construction that interprets the cyclic algorithm. It is instrumental for intuitive understanding of the mechanisms that impair the accuracy of algorithmic converters. In the upper plot of Fig. 5.2, residues are represented as two slanted lines amid the input/output dynamic characteristic comprised between plus and minus V_{ref}. The left and right lines correspond to b of -1 and $+1$, respectively. At the onset, the first residue R_1 (input) is plotted horizontally. The vertical segment it defines represents the second residue R_2. Rotating R_2 by $90°$ and repositioning the segment along the horizontal axis determines the third residue R_3. The same operation is repeated for every other residue. When the last residue is computed, the b_i's hit throughout the conversion are concatenated to yield the output word.

However the Robertson plot is not suitable due to its many vector rotations. A more efficient representation is shown in the lower part of Fig. 5.2. Here, the horizontal and vertical axes are interchanged after every residue. Even residues are plotted horizontally, and odd residues, vertically. The dashed lines represent the residue loci after rotating the axes. First, the residue R_1 is plotted

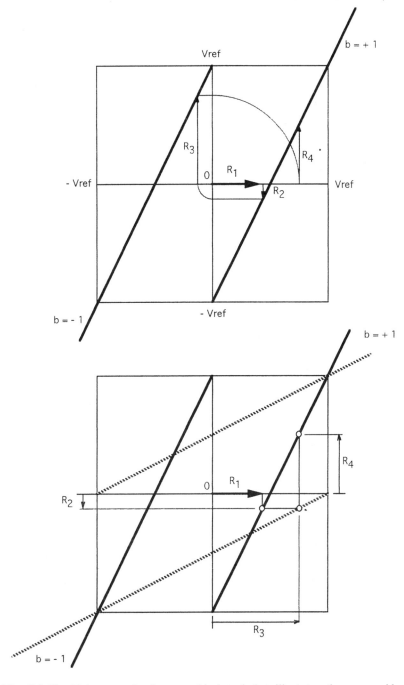

Fig. 5.2 The Robertson plot is a graphical tool that illustrates the way residue calculations perform. The *trajectory* or *signature* shown in the plot below, is obtained when the vertical and horizontal axes are exchanged periodically.

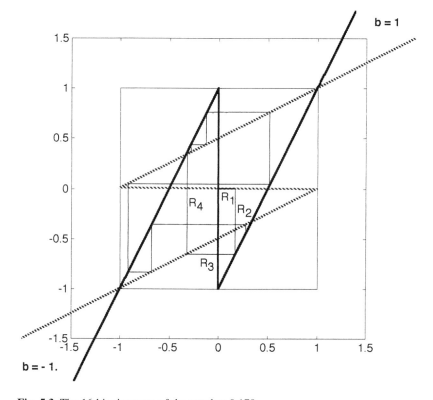

Fig. 5.3 The 16-bit signature of the number 0.170.

horizontally and the vertical residue R_2 is determined as previously cited. The residue R_3 is computed in the new axes; it spans horizontally from the vertical axis until the dashed line that corresponds to b of -1. To compute R_4, the axes are returned to their original positions and the algorithm proceeds like before.

The construction of the Robertson plot is further simplified if one considers only the end-points of the residues. They determine a broken line, called the *trajectory* or *signature* of the number to convert. To devise a signature, vertical and horizontal lines are drawn that alternatively hit a plain or dashed residue line. Once the path is complete, the output word is constructed after concatenating the b_is of the b lines that were hit.

The 16-bit signature of the number 0.170 is shown in Fig. 5.3. The sequence of b's is the same as that listed in the third column of Table 5.1.

When numbers of only a few digits are being converted, limit cycles appear rapidly in the Robertson plot. They illustrate the typical short periodic sequences for such numbers.

5.1.2 Implementation

The implementation of cyclic converters is quite straightforward. It requires few blocks: a sample-and-hold, a comparator and an arithmetic unit. The input signal is stored in the S.H. and its sign is checked against zero by the comparator. The latter outputs a bit b_1 that controls the sign of the reference, subtracted from twice the input. Once the calculation completed, the new residue replaces the old one in the S.H., which is being reconfigured to recycle the other residues.

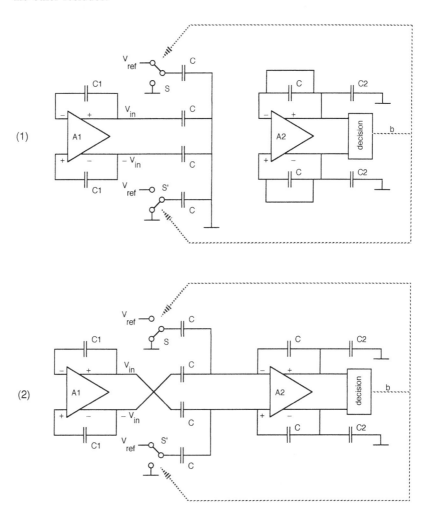

Fig. 5.4 Representation of a switched capacitor algorithmic converter. The sample and computation phases are represented respectively above and below.

Switched capacitor and current copier circuits lend themselves naturally to the implementation of algorithmic converters. In the switched capacitor version, two Op Amps perform the S.H. and arithmetic function. In the current mode implementation, a current copier performs the S.H. operation and summing two identical copies of the stored current does the multiplication by two.

A typical differential switched capacitor algorithmic converter is illustrated in Fig. 5.4 (a differential architecture is preferable because it reduces the impact of substrate noise while doubling the dynamic range). Op Amp A1 performs the sample-and-hold function while A2 does both the comparison and the arithmetic calculation. During the sample mode (a), the residue is stored in A1. The switches S and S' take the positions determined by the bit b issued during the previous cycle. During phase (b), the four capacitors between A1 and A2 transfer their charge to the feedback capacitors of the second Op Amp to perform the arithmetic calculation. Multiplication by two is done by cross-coupling the output terminals of A1. When steady state conditions are met, the decision block checks the sign of the output of A2 to fix the next bit b. After this, the circuit returns to mode (a), and the newly calculated residue is transferred from A2 to A1 by swapping the capacitors C1 and C2.

The current-mode version of the cyclic converter illustrated in Fig. 5.5(b) uses a slightly modified version of the algorithm, known as the single quadrant non-restoring algorithm [Nairn and Salame 1990]. During the sampling phase, the input signal is stored twice in the lower N-type current copiers. The sum is copied in the single P-type current copier above which performs the multiplication by two. The current delivered by the latter is compared with the reference I_{ref}. If larger, b is set to 1; if not, it is set to 0.

The principal sources of errors that impair switched capacitor algorithmic converters are the incorrect multiplication by two, Op Amp offsets and charge injection from MOS switches. The current copiers in the current-mode implementation introduce errors caused by their finite output impedance as well as by charge injection from switches. Proper measures are required to minimize the non-linear distortion from signal dependent charge injection. The injected charge is generally turned into a constant common mode offset voltage that is ignored by the differential architecture (Section 8.2.9). Desaturation of the memory transistors is advocated, for it renders the transconductance of the memory transistor less dependent on the magnitude of the stored current. The unavoidable reduction in output impedance resulting from desaturation is compensated for by cascoding the memory transistor and adding an eventual feedback loop around the current copier (a circuit known as the regulated cascode). Of course, static current mirrors (Fig. 5.5(a)) would elude the charge injection problem but do not represent a good alternative as they are too sensitive to mismatch as illustrated by the poor performances of the 8-bit resolution converter described in Wang (1991). The 14-bit resolution of the converter described in Deval *et al.* (1991) clearly displays the better accuracy of dynamic current copier implementations (Fig. 5.5 (b)), especially

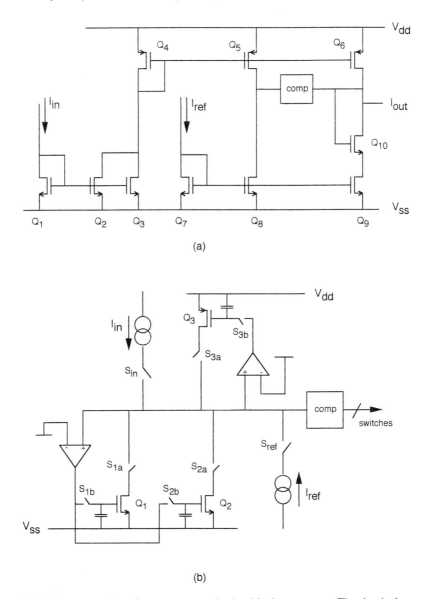

Fig. 5.5 Representation of two current-mode algorithmic converters. The circuit above makes use of static current mirrors while the one below takes advantage of dynamic current copiers.

102 | Algorithmic A to D converters

for mixed mode integrated circuits. However the speed is not better than that of switched capacitor converters. Comparing currents resumes to either probing a voltage across a high impedance node with little but unavoidable parasitic capacitance, or using the summing node of an Op Amp. Unfortunately both have a negative impact on the bandwidth. As a result, current mode converters are generally not faster than voltage mode algorithmic converters (if not slightly inferior).

As stated previously, algorithmic converters are a subclass of a wider class of converters known as multi-bit recycling converters, extensively discussed in Chapter 8. A multibit recycling convertor is represented in Fig. 5.6. The comparator is the single-bit A to D converter that delivers either a plus or a minus one. This signal controls the single-bit D to A converter, which outputs minus or plus $V_{ref}/2$, where V_{ref} is the dynamic range of the converter. The next residue is obtained subtracting the output of the D to A converter from the input, and multiplying the difference by the interstage gain of 2.

5.1.3 Accuracy issues

This section reviews the items that control the accuracy and describes a few correction techniques. The main source of errors comes from the interstage non-ideal gain that produces irreversible degradation of the residues. The smaller the error of the interstage gain, the more bits can be issued before an erroneous bit is generated.

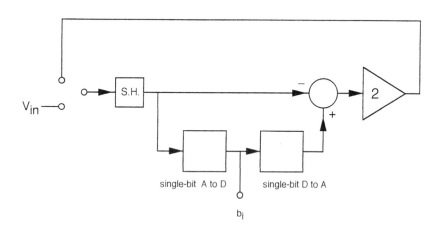

Fig. 5.6 Generalized representation of algorithmic converters.

The cyclic algorithm | 103

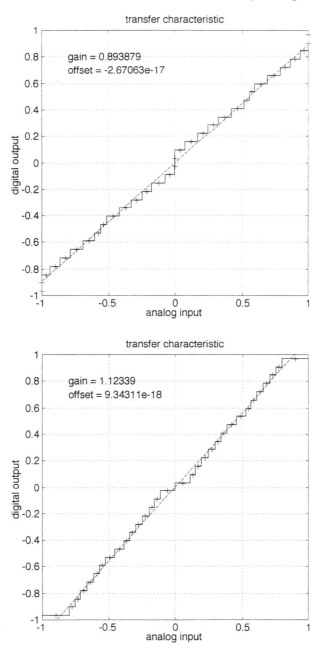

Fig. 5.7 Interstage gain errors corrupt not only the *INL* and *DNL* but also modify the slope of the transfer characteristic.

The interstage gain error

The accuracy of the interstage gain should be the same as that of the converter itself. Matching is a way to take care of this, as long as the resolution does not exceed 9 or 10 bits. When tighter tolerances are contemplated, other more appropriate correction techniques become essential. Figure 5.7 shows the transfer curves of two converters after 6 cycles. Their interstage gains, instead of being 2, are respectively 1.8 and 2.2; no other errors are considered. In the upper figure, two missing codes are spotted near the center, and many *DNL* errors are visible everywhere else. The slopes of the transfer characteristics after linear regressions are respectively 0.894 and 1.123.

Most interstage gains do not suffer from such large deviation since the error is generally much smaller, typically 1 to 0.2 per cent. Thus a detailed examination of the *INL* and *DNL* plots is essential. The two plots of Fig. 5.8 represent the *INL* and *DNL* after 10 cycles of a converter whose interstage gain is 0.1% too large. Both curves are referenced against the linear regression of their transfer characteristics, to overcome the non-ideal overall gain mentioned earlier. Sharp discontinuities are spotted when the *MSB*s change. The largest *INL* reaches one *LSB*, and the *DNL*, one-half *LSB*. If the gain is changed from

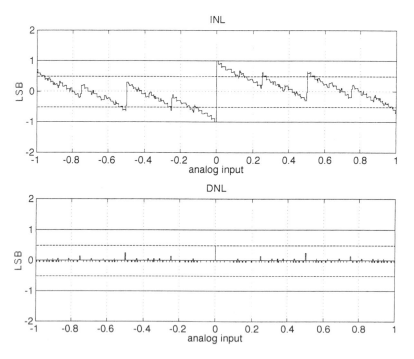

Fig. 5.8 *INL* and *DNL* plots of a 10-bit converter whose interstage gain is equal to 2.002 (no other impairments).

The cyclic algorithm | 105

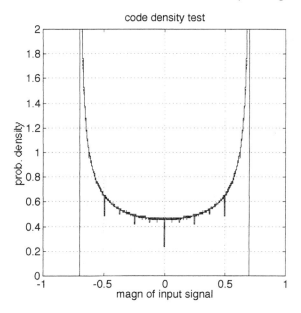

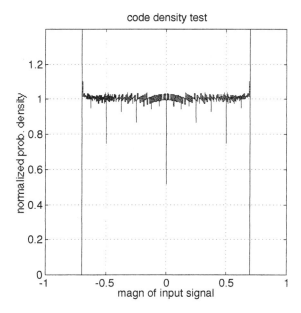

Fig. 5.9 When the interstage gain is smaller than 2, large negative *DNL* errors are generated as illustrated by the Code Density Test.

106 | Algorithmic A to D converters

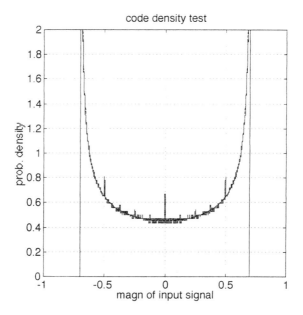

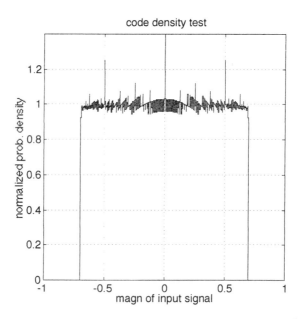

Fig. 5.10 When the interstage gain is large than 2, large positive *DNL* errors are generated as illustrated by the Code Density Test.

2.002 to 1.998, the same *INL* and *DNL* curves are obtained after exchanging the right- and left-hand sides of the plot.

Identical conclusions may be drawn from the code density test. The upper plot shown in Fig. 5.9 represents the outcome of the code density test of a 10-cycle converter whose interstage gain is 1.998. Its input consists of a pure sine wave whose magnitude equals 0.7 V. The lower plot, called the 'normalized probability density', is obtained after referring the code density test to the probability density of the input sine wave. Identical conditions prevail in Fig. 5.10, with the exception of the interstage gain, which is 2.002 instead of 1.998. The probability of occurrence of some code words tends to lessen when the gain is smaller than 2, whereas the opposite applies when the gain is larger.

The signal-to-noise plot versus the magnitude of the input signal shows the impact of the interstage gain error in another way. The two curves shown in Fig. 5.11 represent the *SNR* plots of two converters, both running through 13

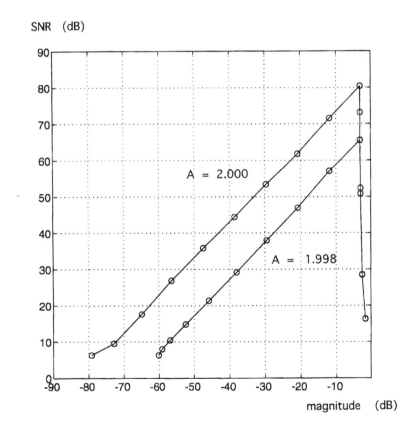

Fig. 5.11 Comparative *SNR* plots of an ideal converter and the same converter after introduction of an interstage gain error of 0.1%.

108 | Algorithmic A to D converters

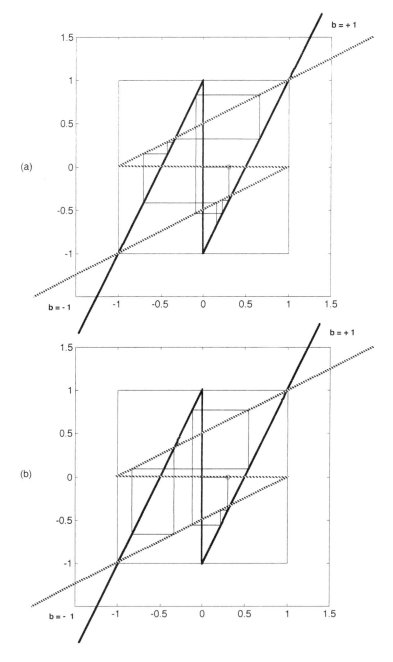

Fig. 5.12 Illustration of the progressive signature degradation caused by the interstage gain error. Above the interstage gain is ideal while below it is 0.1% off.

cycles. The upper figure represents the ideal converter. Extrapolating the *SNR* obtains 80 dB, in good agreement with the expected 13 bit resolution. The lower curve represents the same converter after the interstage gain is affected by 0.1%; the *SNR* reaches no more than 64 dB, or 10.3 bits. This clearly confirms that the accuracy of the interstage gain should be the same as that of the converter itself.

The Robertson plots displayed in Fig. 5.12 illustrates the mechanism that jeopardizes accuracy. The upper plot relates to the ideal converter and the lower, to the same converter after the interstage gain has been increased by 1%. No other sources of errors are considered. The reference voltage, the input signal and the number of cycles are identical, respectively 1 Volt, 0.3072 V and 16 cycles. In the lower plot the signature departs progressively from the ideal as the cycle count increases. A first erroneous bit occurs after the ninth hit, when the '*b*' line with an opposite sign is hit. From there on, the trajectory is irremediably invalidated (the output after the 16th cycle is actually 0.3053 instead of 0.3072).

Gain errors result not only from passive component mismatches but also from Op Amp impairments. A large open loop gain is required to ensure the accuracy of the arithmetic operation. At least 80 to 90 dB are needed to comply with a 10-bit resolution. In addition, the gain must remain large all over the dynamic range of the Op Amp to minimize the distortion that otherwise results from the non-linearity of the transfer characteristic whenever large residues are computed.

Gain error correction techniques

This section reviews three techniques to correct the interstage gain error section: *recirculation of the reference voltage* (Shih and Gray 1986), *ratio independent multiplication by two* (Shi et al. 1983; Li et al. 1984) and *auto-trimming* (Ohara et al. 1987).

The technique known as 'recirculation of the reference voltage' takes advantage of the concurrent transformation of the signal and the reference voltage. The idea is to affect the reference by the same error as the factor two, cycle after cycle. The converter ignores the error under these circumstances. Let us assume the gain A may be expressed as:

$$A = 2 \cdot \left(1 + \frac{\varepsilon}{2}\right) \quad (5.5)$$

where ε is the relative gain error. If the reference voltage V_{ref} is affected by the same error cycle after cycle, the reference after k cycles is defined as:

$$V_{\text{ref},k} = V_{\text{ref}} \cdot \left(1 + \frac{\varepsilon}{2}\right)^k = V_{\text{ref}} \cdot \left(\frac{A}{2}\right)^k \quad (5.6)$$

110 | Algorithmic A to D converters

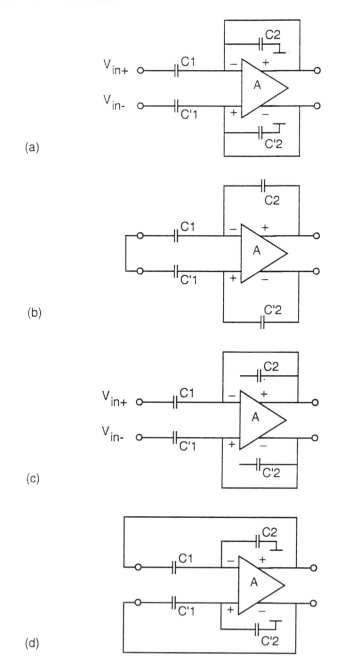

Fig. 5.13 Illustration of the ratio-independent multiplication by two algorithm.

After three consecutive cycles the residue is thus:

$$A\{A(A R_0 - b_1 V_{\text{ref},1}) - b_2 V_{\text{ref},2}\} - b_3 V_{\text{ref},3} \tag{5.7}$$

Rearranging the terms obtains:

$$A^3\left\{R_0 - \left(b_1 \frac{V_{\text{ref},1}}{A} + b_2 \frac{V_{\text{ref},2}}{A^2} + b_3 \frac{V_{\text{ref},3}}{A^3}\right)\right\} \tag{5.8}$$

or

$$A^3\{R_0 - (b_1 2^{-1} + b_2 2^{-2} + b_3 2^{-3})V_{\text{ref}}\} \tag{5.9}$$

Thus the binary fractional expansion obtained is correct regardless of the gain error. However, extra time is needed to perform the conversion since every bit must run the algorithm twice: once for the residue and once for the reference.

The 'ratio independent cyclic analog-to-digital conversion' algorithm proposed by Li et al (1984) contemplates a different approach. Here the multiplication by two does not rely on matching. The multiplication is achieved by storing the same amount of charge twice on the same capacitor. This requires an auxiliary capacitor to provisionally collect the first charge packet while the second is being sampled. The implementation of the algorithm is shown in Fig. 5.13. Under (a), the input is stored for the first time on the capacitors C1 and C′1. Concurrently, the capacitors C2 and C′2 are discharged and the Op Amp feedback loop is shorted. When the second phase (b) starts, the charges across C1 and C′1 are transferred to the auxiliary capacitors C2 and C′2, which close the feedback loop around the Op Amp. The capacitors C1 and C′1 are now ready to store the input data for a second time, and C2 and C′2 kept open (c) to save the first sample. Lastly, during phase (d) the first charge packet is reshuffled to the capacitors C1 and C′1 by closing the feedback loop across the Op Amp. Although the principle appears sound, practice deems the need for extra care to minimize the impact of stray capacitances.

In the technique described by Ohara et al. (1987), the capacitance ratio that defines the interstage gain is trimmed during idle periods. The converter input is the reference voltage V_{ref}. Thus all residues supposedly reproduce the reference. In practice however, due to the interstage gain error, the magnitude of the residues tends to increase or decrease exponentially while the number of cycles increases. This is useful information to estimate the interstage gain error. The later the sequence diverges, the closer the gain approaches the nominal value. Correction is done by adding or subtracting small capacitors that belong to the trim array shown in Fig. 5.14. The idea is to adjust the trim array to lengthen the sequence of uniform residues as much as possible. The operation is repeated several times until no consistent correction can be found. Usually, the limit is set by Op Amp noise.

112 | Algorithmic A to D converters

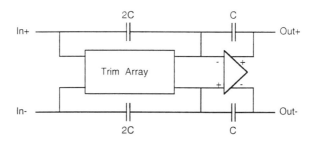

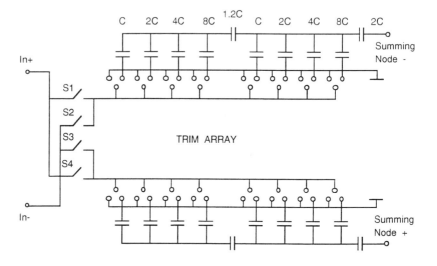

Fig. 5.14 The trimming array used in (Ohara *et al.* 1987) to adjust the interstage gain.

Instead of residues, the converter bit stream output can be used for calibration too. Since the converter nominal output is one, an infinite sequence of identical bits should be expected unless the interstage gain is incorrect. As long as the interstage gain is smaller than two, the output sequence displays bits with variable signs. If the gain is larger than two, the sequence however consists always of plus or minus ones, regardless of the gain error. The calibration fails consequently. The reason for this is determined, contemplating the Robertson plot shown in Fig. 5.15. When the gain is larger than two, b lines cross each other inside the V_{ref} square near the N–E and S–W corners. This creates openings through which the signature leaves the reference square immediately after the first cycle. Once this happens, the signature bounces away against b lines that have the same names. Thus the result is the same as if the interstage gain were correct. This does not occur when the interstage gain is

The cyclic algorithm | 113

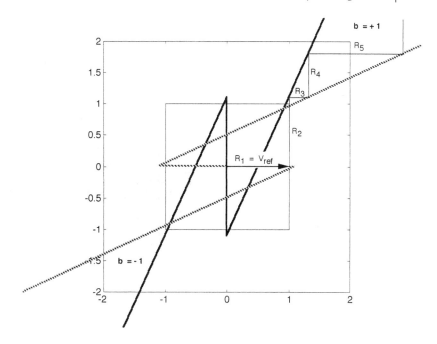

Fig. 5.15 When the interstage gain exceeds 2, the calibration algorithm fails.

smaller than two, for the b lines cross then each other outside the square, as shown in Fig. 5.16. The signature is bound to stay then within the reference square while it departs progressively from the corner. After several b lines with the same names are hit, a line with another name is encountered. The number of hits having the same weight is actually a measure of the interstage gain error. Thus, calibrations based on the output bit stream are legitimate, but only if the gain is less than two. The correction must therefore proceed cautiously when nearing completion. This is not easy. A calibration technique that does not suffer this drawback will be described later.

Effects of offset errors

Let us consider now the impact of the comparator and Op Amp offsets on converter accuracy. The interstage gain is assumed ideal throughout this section, to clearly separate offsets discussion from interstage gain errors. The comparator translates the transition from minus to plus b lines, shown as two little arrows in the upper Robertson plot of Fig. 5.17. The Op Amp offset produces equal shifts of the b lines parallel to the axes as shown in the lower part of the same figure.

114 | Algorithmic A to D converters

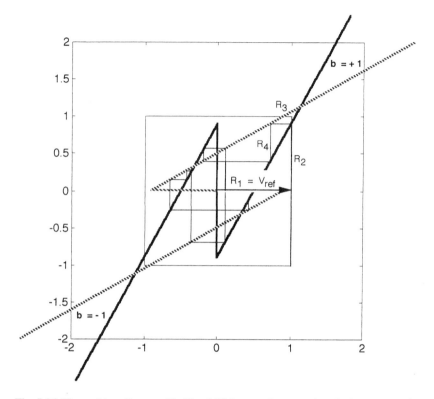

Fig. 5.16 The problem illustrated in Fig. 5.15 does not happen when the interstage gain is smaller than 2.

Let us first consider the comparator offset (upper plot). As long as the trajectory stays away from the two shaded areas outside the reference square, the offsets do not impair accuracy. The signature hits identical b lines and consequently yields the same code words. However once it enters a shaded area, the signature 'escapes' from the reference square to get caught between two b lines with the same names where it bounces forever.

A similar situation is experienced with the Op Amp offset. The offset shifts the plain residue lines in the lower plot of Fig. 5.17 parallel to the vertical axis, whereas the dotted lines are shifted horizontally. This is similar to an equal and opposite shift of the reference square, which is the same as adding the offset to the input. Consequently, as long as the trajectory does not 'escape', the output represents the input increased by the offset.

Hence, unlike the interstage gain error, offsets, whether related to the comparator or the Op Amp, do not introduce non-linearity *as long as the*

The cyclic algorithm | 115

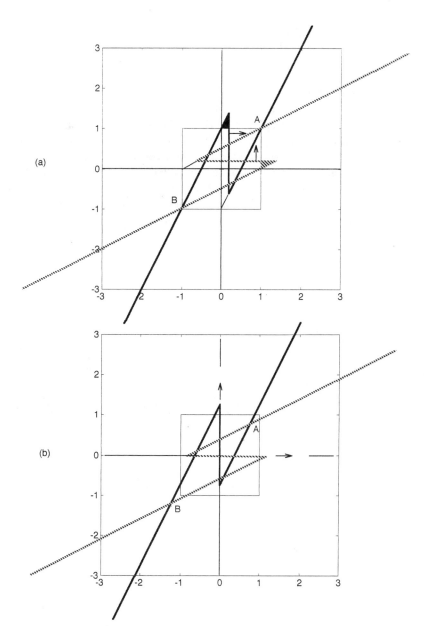

Fig. 5.17 Impact on the Robertson plot of the comparator offset (above) and Op Amp offset (below).

signature remains within the reference square. This is not a very useful conclusion though for 'escapes' are unpredictable by essence. Offset correction techniques like the auto-zero algorithm described in Degrauwe *et al.* (1985) thus seem necessary. The next section shows how to avoid this.

Offset errors correction

Trajectories 'escape' due to saturation of the single-bit A to D and D to A devices shown in Fig. 5.6. The comparator cannot quantize signals larger than plus or minus V_{ref} and the D to A cannot output data exceeding plus and minus $V_{ref}/2$. Thus residues over V_{ref} are necessarily distorted. This is overcome by extending the dynamic ranges of both the A to D and the D to A converters. It means that 4, rather than 2, quantization levels are needed, even though a single bit per cycle is required. These two-bit words change the residue loci in the Robertson plot from those shown in Fig. 5.17 to those of Fig. 5.18. Instead

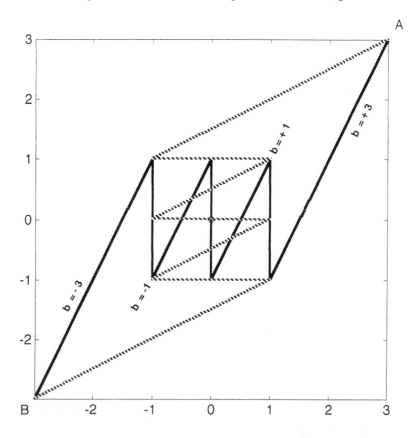

Fig. 5.18 Robertson plot of an ideal converter whose dynamic range has been multiplied by three to counteract distortion caused by offset errors.

of two segments or b lines, we have four b lines named -3, -1, $+1$ and $+3$ V_{ref}, respectively. These are separated by three transitions located at -1, 0 and $+1$. Each b line corresponds to a 2-bit output word and each transition, to one out of the three comparison levels of the 2-bit A to D converter.

Signatures cannot 'escape' for the points A and B beyond which trajectories get lost lie now well beyond reach. Offsets cannot create situations similar to those of Fig. 5.17. The offsets of the A to D converter introduce small variations of the transition positions around their equilibrium positions, and Op Amp offsets resume to small shifts of the center square. The signatures are substantially modified, but the conclusions stated in the previous paragraph are justifiable for they cannot 'escape' anymore. Transition positions consequently do not threaten the reconstructed output code. Neither does the Op Amp offset, which simply adds to the output. The incorrect data from the A to D converter are corrected in fact during the output data reconstruction phase because the least significant bit of every pair overlaps the most significant bit of the pair issued during the next cycle. Consequently incorrect decisions are defeated while the conversion proceeds.

Another problem lies in the accuracy of the D to A converter, since its performance has a direct effect on the output words subtracted from the input data. The *INL* and *DNL* performances of the latter affect indeed the accuracy of the analog data that are recycled. Before the conversion scale was extended, there was no problem because the data outputted by the D to A converter could only take two levels, thus neither the *INL* nor the *DNL* of the D to A converter were concerns. However, this no longer applies with the redundant 2-bit representation. Since the magnitude of the transitions is now essential, matching is required. Alternatives like thermometer-instead of binary-coding may be contemplated as they contribute to improve the overall spectral performances. Another approach that does not require matching is described in Section 5.2.

Noise considerations

Invariance with respect to transition position errors holds even when transitions change during the actual conversion. Since comparator noise can be assimilated to random modifications of the transition positions, the A to D converter noise has no impact on the converter accuracy. Incorrect decisions taken while the conversion proceeds, are cancelled out during the course of the conversion.

The same does not hold true for Op Amp noise. This noise may be assimilated to a variable offset. Since the latter adds a changing error to every residue, small step widths modifications will be induced all over the transfer characteristic. In other words, Op Amp noise affects the converter *DNL*. The problem softens however as far as the $1/f$ noise; very low frequency noise may be assimilated to a constant offset if the sampling frequency exceeds the corner frequency.

118 | Algorithmic A to D converters

5.2 The RSD algorithm

Redundant Signed Digit (RSD) converters (Ginetti 1992; Gani and Gray 1990) take advantage of the *SRT* algorithm (after Sweeny, Robertson and Tocher) described in Hwang (1979). The algorithm was known long before cyclic converters were being implemented. It makes use of single ternary digits instead of 2-bit words. Thus the output of the D to A converter takes three instead of four values and the A to D converter has two, instead of three, discrimination levels. For the remainder, the similarity with the 2-bit redundant implementation above is obvious.

The plot represented in Fig. 5.19 shows what happens when cyclic converters are turned progressively into RSD converters. The thick curve in the middle represents the transfer characteristic of a converter with a single-bit

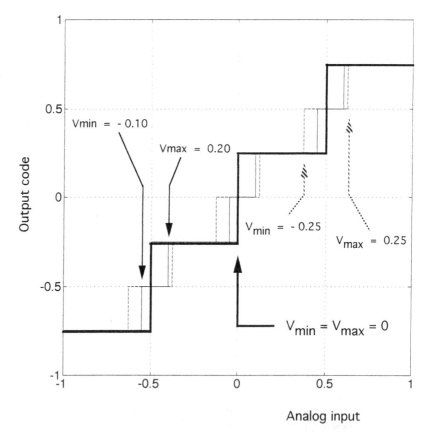

Fig. 5.19 The additional bit inherent to RSD converters results from splitting the steps of the non-restoring algorithmic converter.

quantizer running two cycles. The other curves represent characteristics of the same converter after the zero threshold of the single-bit comparator is split into two transition levels, V_{min} and V_{max}. The transfer characteristic in thin plain lines corresponds to transitions equal to −0.10 and +0.20, and the dashed lines correspond to −0.25 and +0.25 (V_{ref} is assumed to be equal to one). Each step of the original transfer characteristic is divided into two sub-steps whose widths vary proportionally with the magnitude of the transitions. When these equal minus and plus 0.25, the transfer characteristic looks like that of an ideal 3-bit converter with 7 instead of 8 levels. The missing level that could turn the RSD converter into a true 3-bit binary converter is a consequence of the particular transfer characteristic of RSD converters; these exhibit a zero level that is ignored in the other converters. If more than two cycles are considered, the resolution is almost twice that of the original converter.

RSD and algorithmic converters that use 2-bit quantizers share identical properties regarding their insensitivity to offsets. The extended dynamic range inherent to the three state digits leaves room to perform code corrections like in the previous converter, after its resolution was extended from one to two bits. Offsets related to the A to D converter translate into transition position errors, which alter the string of digits delivered by the converter, but have no effect on the reconstructed output data. Since the tolerances on the transition levels can be as large as plus or minus $V_{ref}/4$, simple yet inaccurate devices like Schmitt triggers, suffice to implement the A to D conversion. The result is not only substantial savings in area but also higher speed of the A to D front-end. The Op Amp offset neither does not affect linearity, its magnitude is simply added to the output data producing a concurrent global shift of the transfer characteristic, easily removed through the use of the compensation algorithm described in Ginetti (1992).

The D to A converter linearity is no longer a problem because unit-steps with equal and opposite signs are easily implemented in switched capacitor circuits. Thus the enactment from four to three levels is of considerable interest. The interstage gain is the only remaining problem. The calibration should proceed along the lines described earlier. But the condition that the gain be smaller than two is no more the case, as offset errors no longer interfere with the calibration procedure.

5.2.1 The Robertson plot

Extension of the Robertson plot to RSD converters is very straightforward. The residue locus consists of three sections. In Fig. 5.20 the signature of the trial number 0.9123 is displayed for an ideal and a non-ideal converter considering 16 cycles. The insterstage gain is assumed ideal in both converters, but the transition positions and offsets differ. The ideal situation is depicted in the

120 | Algorithmic A to D converters

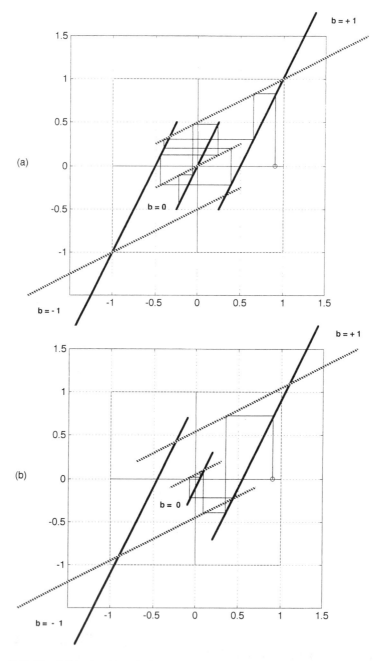

Fig. 5.20 The RSD algorithm is insensitive to transition position errors. The ideal converter (above) and non-ideal converter (below) Robertson plots display completely different signatures but the reconstructed output data are the same.

upper part. In the lower part, the transitions positions change respectively from −0.25 to −0.10 and from +0.25 to +0.20 while the Op Amp offset is increased from 0 to −0.10. Once the Op Amp offset is subtracted from the input, the reconstructed output data yields relative errors of less than 0.01% in both cases.

In the example of Fig. 5.20, the only region where signatures might eventually escape is the S–W corner. Since the width of the escape window is determined by the Op Amp offset, the chances of losing signatures are restricted to very small limits. The most cumbersome situation however occurs during the calibration of the interstage gain, for the input equals then the reference. A way to avoid the problem is to perform calibration for both plus and minus reference values. The only sequence to consider for the calibration is that containing different signs.

5.2.2 Floating point RSD converters

In the plot of Fig. 5.11, two *SNR* curves were compared to show the impact of the interstage gain error on the accuracy of algorithmic converters. When the same test is performed on an RSD converter, a few interesting features are highlighted (Grisoni *et al.* 1997). First, the maximum *SNR* of the RSD device is always 6 dB above the non-RSD converter owing to the additional bit from the RSD converter. Second, instead of dropping proportionally to the input signal, the SNR remains more or less constant as the magnitude decreases until the *SNR* locus of the ideal converter is reached (Fig. 5.21). This behavior is reminiscent of companders or CODECs and is similar to the floating point representation of digital numbers.

The reason is easily understood by comparing the plots of Fig. 5.22, which illustrate the signatures of two ideal converters with very small input signals. In the single-bit A to D and D to A implementation, the first hit with a 'b' line occurs along the vertical axis, near the transition. From there, the signature is bent either to the N–E or to the S–W corner where it bounces repeatedly between lines having the same names until it spreads over the entire reference square. Thus, the sequence of output data consists of a first bit followed by many opposite sign bits. These progressively equilibrate the unduly large magnitude commanded by the *MSB* to keep the reconstructed number close to zero. In the RSD converter, things happen differently. The signature slowly builds up in the center where it repeatedly hits '0' lines before finally spreading around. In this way, the reconstructed output word consists of a long sequence of zeros before other data are acknowledged. Now consider the impact of an interstage gain error. In the first converter, the gain error takes the 'b' lines intercepts away from the N–E and S–W corners along a diagonal of the reference square. Consequently, less bounces occur before the signature spreads all over the square. The sequence of bits that supposedly

122 | Algorithmic A to D converters

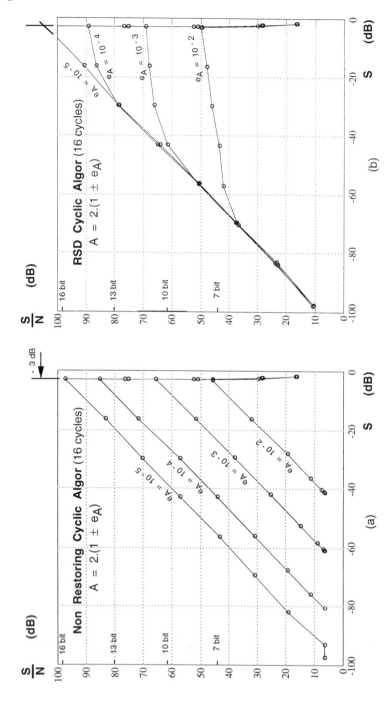

Fig. 5.21 Comparative *SNR* plots of the non-restoring converter (left) and an RSD converter (right).

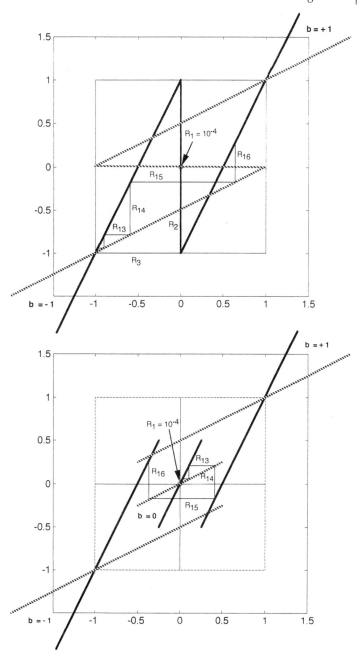

Fig. 5.22 The improved *SNR* behavior of RSD converters (below) compared to non-restoring converters (above) for small signals is a consequence of the existence of the zero b line proper to RSD converters.

compensate the *MSB* is shortened. In the RSD implementation, the trajectory still exhibits a sequence of 0 digits, for the gain error only changes the slopes of the '0' lines. The length of the sequence is slightly less of course, and this explains the small slope deviation from horizontal of the *SNR* in the middle. Note however that this only holds true if the Op Amp offset is not considered. The latter shifts the intersection of the zero 'b' lines away from the origin and creates a situation that shortens the length of zeroes as well. Extended simulations show however that the impact of Op Amp offsets is less worrying than could be expected. The effect is similar but less than that of noise; for instance, an offset of 5 mV produces the same result as Op Amp noise as large as 100 µV.

The *SNR* curves plotted in Fig. 5.23 show the combined effects of the interstage gain error and wide-band noise from the Op Amp. We start from an ideal RSD converter running 20 cycles. In this device, the *SNR* follows the dotted line called the 'theoretical limit'. In the 'real' converter, noise shifts the *SNR* plot to the right while the interstage gain error sets a limit to the resolvable resolution. It is interesting to note that CCITT specifications can easily be met even with a 1% mismatch and 100 µV noise.

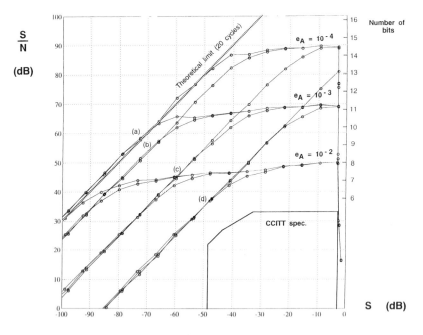

Fig. 5.23 *SNR* plots of an RSD converter running through 20 cycles in presence of an interstage gain error and Op Amp noise. The interstage gain error takes three values from 0.01, 0.1 to 1% while the standard deviation of the Gaussian wide band noise at the output of the Op Amp varies from 0 (labelled a) to 1 (b), 10 (c) and 100 µV (d).

5.2.3 Current mode RSD algorithmic converters

Current-mode converters use a slightly modified version of the RSD algorithm that avoids the sign changes of the currents delivered by the current copiers (Macq and Jespers 1994). The input is compared with two currents, I_{min} and I_{max}, respectively equal to 3/4 V_{ref} and 5/4 V_{ref} (Fig. 5.24). These currents are the counterparts of the voltage comparison levels peculiar to previous RSD converters. If the input current lies below the smallest of these currents, the

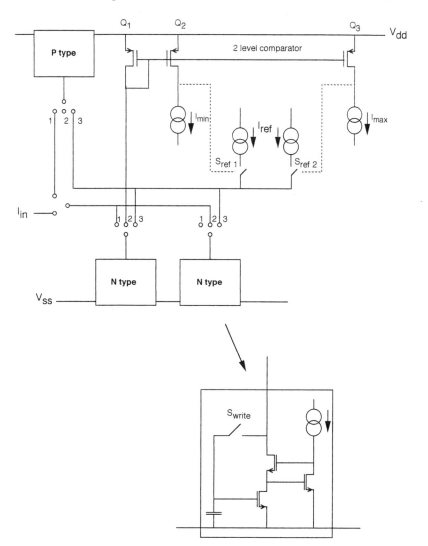

Fig. 5.24 A current-mode RSD converter.

input is simply multiplied by two. If it lies beyond the largest, I_{ref} is subtracted twice from the doubled input; if in between, a single Iref is subtracted from twice the input. The output bit b takes the values of 0, 2 or 1, respectively.

A block diagram representing a current-mode RSD converter is shown in Fig. 5.24. It generalizes the circuit shown in Fig. 5.5. The main difference lies in the comparison. During phase '1', the input current is stored in the first of the two N-type current copiers below. When the second sampling occurs in phase '2', the copy stored in the first copier is compared to I_{min} and I_{max}, using the static P-channel current mirrors above. This sets the positions of the switches Sref1 and Sref2. During phase '3', the sum of the currents stored in the two N-type current copiers are added to the reference currents combination selected according to the outcome of the comparison that took place during phase '2'. The result is stored in the P-type current mirror above. The same scenario repeats itself from there on, the input being the current delivered by the upper current mirror. As in voltage mode implementations, errors impairing the comparison levels I_{max} and I_{min} are corrected automatically, thus static current mirrors may used for the comparisons, regardless of their potentially large mismatch errors.

Rampfunction converters 6

Integrated rampfunction converters, though simple and sound, have never received much attention. They use straightforward architectures that exploit the linear relation between magnitude and time typical of integrators. Their linearity can be very high. A discrete Miller integrator easily exceeds 18 bits and requires fewer components. D to A ramp converters require nothing more than a voltage or a current ramp running through a number of clock cycles that is fixed according to the word to convert. Once the ramp stops, its magnitude represents the analog counterpart of the digital input. Similarly A to D converters resume to clock cycle counts that measure the time elapsed before a ramp meets the unknown input analog signal. However, the large linear resistors and capacitors required by these converters cannot be integrated and their conversion times are too long moreover. For instance, a 16-bit converter needs more than 65 000 clock cycles to reach full scale. Thus, a clock of nearly 3 GHz would be required to convert 44 kHz stereo 16-bit audio words. Rampfunction converters cannot cope with such speeds.

6.1 The dual slope A to D converter

The dual-slope converter is a simple high resolution D to A converter, commonly used in instrumentation devices, whose principle has been known since the use of discrete component digital voltmeters. It consists of a Miller integrator, a comparator and a counter. Instead of a single ramp, a double ramp as illustrated in Fig. 6.1 is generally used to facilitate zeroing and calibration. First the unknown input is integrated during a constant time T_1, fixed by a number of clock counts. Then the integrator is switched from the unknown input V_{in} to the reference V_{ref} whose sign must be opposite. While the integrator swings back, the clock count resumes until the integrator output reaches its initial state again. Thus, if T_1 and T_2 are respectively the clock counts during the two phases, and V_m the integrator output at the end of phase one, the following equations can be obtained:

$$V_m = \frac{V_{in}}{RC} \cdot T_1$$

and (6.1)

$$V_m = \frac{V_{ref}}{RC} \cdot T_2$$

128 | Rampfunction converters

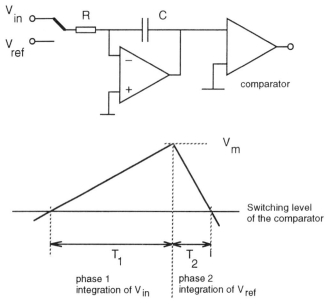

Fig. 6.1 The dual-slope A to D converter.

thus:

$$V_{in} = \frac{T_2}{T_1} \cdot V_{ref} \tag{6.2}$$

Equation 6.2 shows why double integration is preferable to single: the converter is intrinsically insensitive to the actual RC time constant. If the time T_1. is monitored moreover by a number of clock pulses that is a power of ten, the counter output is automatically the decimal counterpart of the input voltage. Another interesting feature is the insensitivity to the comparator's offset, since both T_1 and T_2 are defined with respect to the same indifferent reference. A high resolution is achievable if the linearity of the resistance and the capacitance meet strict specifications. Another interesting feature lies in the fact that the input signal is being integrated. Periodic perturbations superposed on the input may be ignored if the integration time T_2 encompasses entire numbers of periods. In other words, the dual-slope converter operates as if it were preceded by a comb-filter. One may take advantage of this property to make the converter robust to mains-induced noise. An integrated CMOS version of a dual-slope converter is described in Musa (1976). The resistance and capacitor are external elements.

Because the input impedance of the integrator is not infinite, a buffer front-end is usually put put in front of the converter. Eventually this buffer is combined with a S.H. Whichever, the unity gain buffer and the integrator require two Op

Amps. Their combined offsets must be canceled out. A possible strategy is illustrated in Fig. 6.2. In the upper part of the figure, the capacitor Cc senses the integrator offset with respect to the buffer offset by shorting the feedback loop around the integrator. Below, the capacitor is put in series with the buffer once it is turned around. Since the buffer output now replicates the integrator offset, there is no current flow through the resistance: the integrator is zeroed. The same procedure is implemented in the more elaborate sequence shown in Fig. 6.3. In phase 1, zero sensing is achieved like above. During phase 2, the reference voltage V_{ref} is integrated until the integrator output reaches the comparator switching level. This happens after the time T_o which is stored (the existence of a systematic offset between the auto-zeroed output of the integrator and the comparator switching level is assumed). Phase 3 is a replica of phase 1 to re-establish the conditions prevailing before conversion. The input voltage is integrated during phase 4, followed by integration of the reference voltage. During phase 5, the counter is decreased until it reaches zero. This automatically subtracts T_o from the time interval, separating the end of phase 4 from the moment the integrator reaches the switching level of the comparator. The output then the represents the input data.

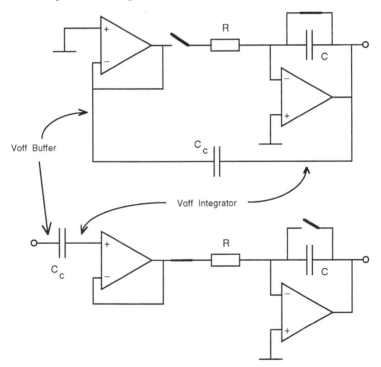

Fig. 6.2 The auto-zero procedure used in dual-slope converters compensates the offsets of both the input buffer and the integrator.

130 | Rampfunction converters

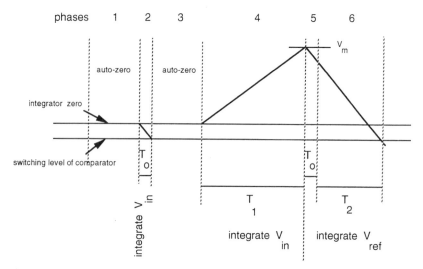

Fig. 6.3 The automatic compensation of offsets in this dual-slope converter is done according to a calibration procedure that takes 6 phases.

The simplicity of the dual-slope A to D converter, combined with its potential for low-power consumption and high accuracy (4.5 digits or 15 bits) make this device suitable for portable digital multimeters. However external RC components are required to yield comparable performances, a condition that is waived in the Delta–Sigma voltmeter described in Section 7.6.

6.2 Improving the speed of rampfunction D to A converters

Among the attempts made to shorten the conversion time of rampfunction converters, double and triple ramp converters deserve mention. The concept behind the use of these converters is to concatenate ramps with multiple slopes to shorten the conversion time. For instance, a fast 10-bit D to A rampconverter can be implemented with two ramps: a slow one for the *LSBs* and a fast one for the *MSBs*. The *MSB* field is first converted with a ramp that has a very steep slope. Then the *LSB* field is converted with a ramp whose slope is 32 times smaller than the previous one. The final output voltage when superposed is the correct 10-bit analog representation of the input. Compared with the single rampfunction, the time to reach the F.S. now equals the sum of the times needed by each of the two ramps to reach half F.S. If T is the clock time, a 10-bit single ramp converter requires 2^{10} time slots whereas the double slope converter needs only two 2^5 time slots. The multiple slope converter is thus 16 times faster. Larger ratios are possible when more than two ramps are considered.

One of the first double ramp audio D to A devices intended for 16-bit resolution converters is described in Kayanuma et al. (1981). The 16-bit input code word is divided into two 8-bit fields. The *MSB* field controls a rampfunction 256 times faster than that controlling the *LSB* field. Only 512 clock pulses are required to reach the full scale. The clock frequency is now as low as 30 MHz instead of the 3 GHz mentioned above. The idea is elegant, but reality is somewhat different. The ratio between the two current sources has to be precise and remains so over time. High speed logic must be associated with the most significant count. Indeed, the accuracy of the stop-voltage must be controlled with the same precision as the single slope converter. This implies that the largest of the two current sources be switched with a precison of a few picoseconds. Therefore, a significant part of the logic consists of fast bipolar ECL circuits that consume a lot of power.

6.3 A charge rampfunction converter

The integration of charge packets is amenable to the implementation of more efficient rampfunction converters whose speed versus accuracy challenge is more acceptable. In a switched capacitor network, this is a matter of Op Amp performance.

In the D to A converter described in Pelgrom and Roorda (1988) a triple ramp charge converter is presented that uses charge packets summation. Only the first two Op Amps are shown in the upper part of Fig. 6.4. The N bit input word is divided into two fields: a P bits *LSB* field and an (N-P) bits *MSB* field. The input word D is then given by:

$$D = [LSB] + [MSB] \cdot 2^P \qquad (6.3)$$

where brackets represent the numerical counterparts of the *LSB* and *MSB* members. The conversion follows the five steps shown in plot (b):

1. both Op Amps are reset,

2. a single packet charge is fed to Op Amp 'A',

3. Op Amp 'B' integrates [*LSB*] times the output of Op Amp 'A',

4. ($2^P - 1$) supplementary charge packets are fed to Op Amp 'A', to reach the full scale output,

5. the output of Op Amp 'A' is integrated [*MSB*] times by Op Amp 'B'.

The principle is similar to that of the previous double ramp converter, despite the fact that the *LSB* field is activated before the *MSB* field.

132 | Rampfunction converters

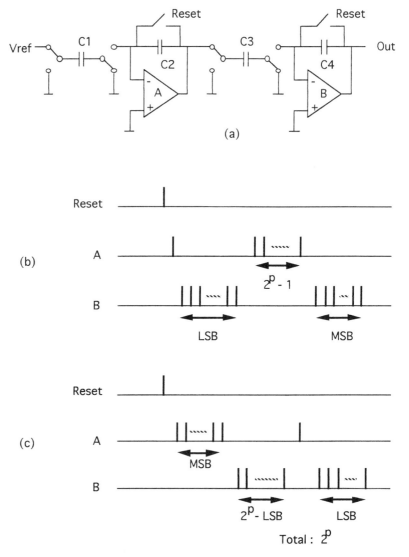

Fig. 6.4 Charge ramp D to A converters produce strong *DNL* errors that are the consequence of the offset of the second Op Amp The *DNL* error is changed in a gain error if the sequence suggested in (c) is adopted.

The drawback of this converter is due to the inevitable offsets of the Op Amps. The offset of the first Op Amp changes the size of the first charge packets. Since this is a constant error, it introduces a gain error that is irrelevant in applications like Hi-Fi audio. The offset of Op Amp 'B' is more awkward for it affects the *DNL* very severely and may cause non-monotonicity. Unless this offset is subtracted from the output voltage of the first Op Amp, the charge stored on C3 is not proportional to the number of charge packets transmitted by the first Op Amp. Consequently there is a large *DNL* error every time a new transfer takes place in the first integrator.

An interesting solution that obtains 15 or 16-bit accuracy is shown in (c). The idea is to modify the switching sequence in such a way that the number of charge transfers of Op Amp 'B' remains constant. The offset needs not to be compensated for its contribution is always the same regardless of the data being transferred. Monotonicity is recovered this way at the expense of an additional gain error. The conversion algorithm that performs this is given by:

$$D = (2^P - [LSB]) \cdot [MSB] + ([MSB] + 1) \cdot [LSB] \qquad (6.4)$$

The total number of charge transfers controlled by Op Amp 'B' is now constant and equal to 2^P, as illustrated in the lower part of Fig. 6.4. The conversion proceeds as follows:

1. Both Op Amps are reset like above.

2. [*MSB*] charge packets are integrated in Op Amp 'A'.

3. The output of Op Amp 'A' is transferred to the output of Op Amp 'B', $(2^P - [LSB])$ times.

4. A single additional charge packet is added to the output of Op Amp 'A'.

5. The new output data of Op Amp 'A' are fed to the output of Op Amp 'B', [*LSB*] times.

A 15-bit converter operating at 44 kHz sampling rate has been implemented according to the above algorithm. The input data are split in three equal 5-bit fields. Only 2 times 2×2^5 or 128 clock pulses are needed for every 20 μs time frame. The clock frequency is thus 6.4 MHz. The actual performances depend on the settling time, the finite gain and non-linearity of the Op Amp. A minimum gain of 80 dB is required to stay within the *INL* specifications. A signal-to-noise ratio of 90 dB is obtained if less than 1 ns clock jitter is achieved. With a transition frequency of 6 MHz and a non-dominant pole of the Op Amp at 30 MHz, the relative amounts of charge that are not transferred per cycle are only 0.05% after 100 ns. Although the principle achieves both accuracy and speed, it has not gained wide attention because more appealing implementations like Delta–Sigma converters achieve similar performances with less severe requirements as far as technology is concerned.

Delta–Sigma converters 7

Progress in the semi-conductor industry has been driven almost exclusively by the prospects of digital applications. State of the art technologies aim towards ever-smaller dimensions to put more transistors on the same chip and make computers run faster. The gate lengths currently range from 0.7 to 0.5 micron, while the objective of more advanced fabrication processes is 0.18 micron. Given the present fabrication yields and packing densities, more complex systems can be integrated. Concurrently the supply voltages are lowered from 5 to 3.3 V to limit the power consumption and 1.2 V supplies are expected in the near future.

Analog integrated circuits take advantage of the improved performance offered by digital technologies. Most of the recent developments in the area of telecommunication integrated circuits would not have been possible without the unprecedented performances achieved with short channel transistors. However, there are some drawbacks. Smaller channel lengths reduce the gain of the transistors, because their output impedance decreases. Abatement of the supply voltage restricts the dynamic range whereas the drain-to-source voltage in the saturation region does not scale down. Hence, the signal-to-noise ratio gets worse; all the more smaller gate areas imply larger $1/f$ noise contributions. In other words, state of the art digital technologies do not ease analog circuit design.

Delta–Sigma converters are an interesting alternative. They exchange the loss of accuracy inherent to analog circuits imbedded in digital technologies for faster signal processing and more digital circuitry. What is lost in magnitude is compensated by resolution in time. Hence, they capitalize the speed of analog circuits and the accuracy of digital circuits.

Many papers deal with the theory, design and simulation of Delta–Sigma converters. Two books are exclusively devoted to them. The first (Candy and Temes 1992) is a collection of key references published in specialized journals like the JSSC. The most comprehensive work published on Delta–Sigma converters is the book by Norsworthy *et al.* (1997). The books by van de Plassche (1994) and Huijsing *et al.* (1993) contain sections devoted to noise shapers and Delta–Sigma modulators. The design of imbedded Delta–Sigma converters intended for digital signal processing is dealt with by Fr. Op't Eynde *et al.* (1993).

What makes Delta–Sigma converters so different from all other converters is the use of stochastic processing to perform conversion. For that reason, this chapter starts with a short review regarding quantization noise and the signal-to-noise ratio (*SNR*) concept.

7.1 Quantization noise

Quantization noise is the difference between analog and digital data, once the latter are reconfigured in the analog domain. It is the difference between the input and output of two converters in cascade: an A to D and a D to A. It is similar to the residue considered in Chapter 5. In the lower part of Fig. 7.1, the quantization noise of a perfect quantizer is shown as the sawtooth shaped waveform below. As long as the input lies within the dynamic range of the quantizer, the magnitude of the quantization noise is restricted to a window Δ located between plus and minus one-half *LSB*. If we assume that the noise is uniformly distributed over the window, its probability density is defined as:

$$P(y) = \frac{1}{\Delta} \tag{7.1}$$

The concept of power is currently associated to quantization noise according to:

$$e_{rms}^2 = \int_{-\frac{\Delta}{2}}^{\frac{\Delta}{2}} y^2 P(y) dy = \frac{\Delta^2}{12} \tag{7.2}$$

The root square of the above expression yields the so-called quantization noise voltage:

$$e_{rms} = \frac{\Delta}{\sqrt{12}} \tag{7.3}$$

The probability density is assumed to be constant. Figure 7.2 displays the 3-bit (eight levels) quantized transform of a continuous pure sine wave and the corresponding quantization noise 'y' in the middle, which clearly underlines the shortcomings of this statement. The probability density exhibits sharp peaks near the edges of the window. Taking the sine wave probability density into account would be more correct, but would complicate the computations and is its usefulness is questionable.

Once the resolution exceeds 6 bits, the quantization noise appears increasingly random. The graph of Fig. 7.3 shows the quantization noise of a 10-bit instead of 3-bit quantizer considering the same input as above (note that the error is multiplied by 200 to visualize the noise). The plot sustains the increasing validity of the constant probability density assumption, designated as the 'busy signal' approximation.

136 | Delta–Sigma converters

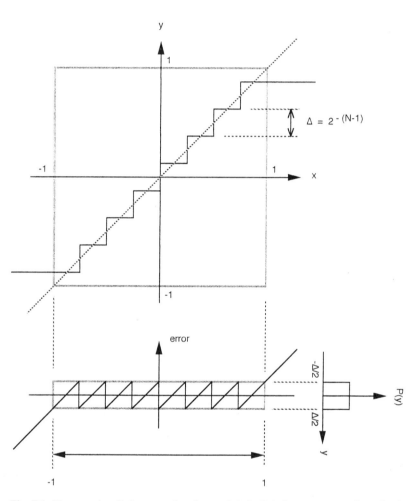

Fig. 7.1 The quantizer links up analog data and their digital counterparts. Quantization noise is the difference between the input and output after reconfiguration of the latter in the analog domain.

Quantization noise is assimilated generally to white noise, thus its power spectral density should be constant. Since the quantizer is a sampled data system, the noise spectrum extends from minus to plus half the sampling

Quantization noise | 137

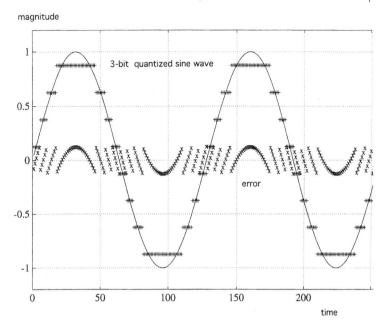

Fig. 7.2 The quantization noise probability density of this coarse 3-bit (8 levels) quantizer is far from being uniformly distributed.

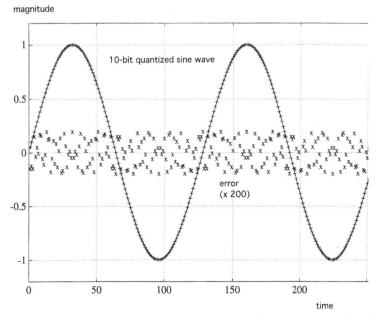

Fig. 7.3 Compared with Fig. 7.2, the random character of the quantization noise of this 10-bit converted sine-wave is much more pronounced (the error vertical scale is multiplied by 200).

frequency f_s while being repeated at every multiple of the sampling frequency. The noise spectral density $S(e_{rms}^2)$ is easily derived from the noise power since its integral over the range from minus to plus $f_s/2$ must represent the total noise power. Thus:

$$S(e_{rms}^2) = e_{rms}^2 \cdot \frac{2}{f_s} = \frac{\Delta^2}{6f_s} \qquad (7.4)$$

because

$$\int_0^{\frac{f_s}{2}} S(e_{rms}^2) df = e_{rms}^2 \qquad (7.5)$$

The integration from zero to $f_s/2$ is justified by the fact that the noise power density is multiplied by two.

The spectra shown in Fig. 7.4 show that these assumptions are relatively consistent with reality, in spite of the fact that the probability density is not truly a constant. All the plots shown in this figure represent fast Fourier transforms (*fft*) of an F.S. quantized sine wave, whose resolution varies from 4 to 16 bits in 3-bit steps. Only half of the 1024 *fft* points are plotted to mimic the frequency axis from DC to $f_s/2$. Both the noise power density and total

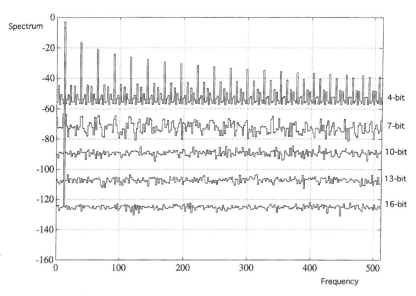

Fig. 7.4 *fft* transforms of a same sinusoidal signal after 4, 7, 10, 13 and 16-bit quantization. The periodic character of the quantization noise vanishes as the quantization granularity gets finer.

noise power derived from the graph are in good agreement with expressions (7.2) and (7.3). When the number of bits gets small, a large number of spectral rays appear. These underline the fact that the probability density of poorly quantized data is far from evenly distributed over the Δ window.

7.2 The signal-to-noise ratio

This section evaluates the signal-to-noise ratio (*SNR*) of the ideal N-bit quantized sine wave. Since the quantization noise voltage is already known from expression (7.3), only the sine wave rms magnitude remains to be evaluated. The rms voltage of an F.S. sine wave is equal to the ratio of the dynamic range $2^N \Delta$ divided by $2\sqrt{2}$. Thus one has:

$$SNR = \frac{\frac{2^N \Delta}{2\sqrt{2}}}{\frac{\Delta}{\sqrt{12}}} = 2^N \sqrt{1.5} \qquad (7.6)$$

Taking the log of the above equation obtains the expression used in Chapter 1:

$$20 \log_{10}(SNR) = 20[N \cdot \log_{10}(2) + 0.5 \cdot \log_{10}(1.5)]$$
or $\qquad (7.7)$
$$(SNR)_{dB} = 6N + 1.8$$

This expression links the *SNR* to the number of bits. Let us consider two examples: the *SNR* of an ideal 16-bit converter is 98.6 dB and that of an 8-bit converter, 49.6 dB. Inversely, once the *SNR* is known, the resolution is found. This statement is the basis of the resolution evaluation method that was presented in Section 1.5.1.

7.3 Increasing the SNR to improve resolution

The Delta–Sigma converter uses *oversampling* and *noise shaping* techniques to lower the quantization noise and consequently improve the resolution.

Oversampling occurs whenever a signal is being sampled at a frequency larger than twice its bandwidth, the so-called *baseband* f_o. According to the Nyquist theorem, the higher sampling rate does not add information to the sampled signal. It is not only inefficient but wrong since the spectrum is unnecessarily widened. Yet, increasing the bandwidth obtains positive results. Consider a quantized signal. Since the quantization noise power remains unchanged regardless of the sampling rate, the noise power density must go down as the spectrum widens. If we restrict the bandwidth of the oversampled

140 | Delta–Sigma converters

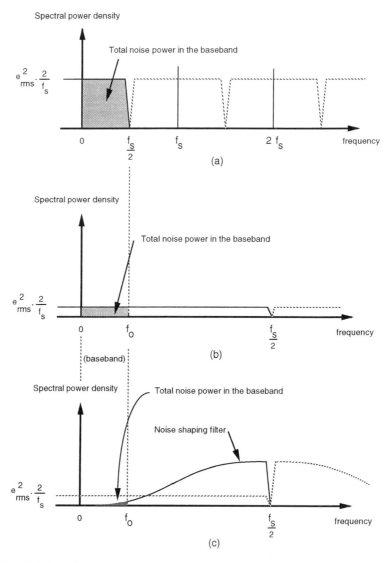

Fig. 7.5 Quantization noise spectral distribution of Nyquist sampled data (a), oversampled data (b) and oversampled plus noise shaped data (c).

signal to the signal baseband no information will be lost, but a lower noise power will be collected in the baseband.

This mechanism is shown in Fig. 7.5. Under (a), a signal is being sampled at the Nyquist rate. The bandwidth of the quantization noise spectral density encompasses the baseband f_o exactly. Under (b), the sampling frequency is multiplied by four. The bandwidth of the quantization noise density is now four times larger, whereas the magnitude is divided by four. If the bandwidth of the oversampled signal is restricted to the baseband, the quantization noise power in the baseband is four times smaller. This is the same as replacing the upper integration limit in eqn (7.5) with f_o:

$$e_0^2 = \int_0^{f_o} S(e_{rms}^2)df = e_{rms}^2 \frac{2f_o}{f_s} = \frac{\Delta^2}{12}\frac{1}{OSR} \qquad (7.8)$$

Thus the noise voltage is defined as:

$$e_0 = e_{rms}\sqrt{\frac{2f_o}{f_s}} = \frac{e_{rms}}{OSR} \qquad (7.9)$$

Throughout the rest of this chapter, the oversampling rate $f_s/2f_o$ will be referred to as *OSR*.

The *SNR* improvement resulting from oversampling is shown in Fig. 7.5 by comparing the gray areas of the a b plots. Since the noise power in the baseband is divided by four, the *SNR* improves by 6 dB, which is the same as adding one bit to the quantized signal. More than one bit is also feasible, but the *OSR* rapidly becomes very large. An *SNR* improvement of 10 bits implies that the sampling frequency be one million times larger than the Nyquist frequency. This is too much unless the signal baseband does not exceed 100 Hz. To improve the *SNR* without compromising the bandwidth, the benefit obtained from oversampling is supplemented by a filtering operation that shifts part of the noise to high frequency, leaving less noise in the baseband. This is called *noise shaping*, and is illustrated in Fig. 7.5(c).

Noise shaping is achieved by putting a feedback loop around the quantizer as shown in Fig. 7.6. The loop controls the quantizer input in such a way that the output tracks the input signal as closely as possible. The role of the low-pass high gain filter controlling the quantizer is to minimize the averaged departures between the input signal and its quantized representation. The loop cannot affect of course the heights of the steps outputted by the quantizer, but tends to modulate the steps of the quantized signal to try to average out the difference between the x and y signals. Since the quantized signal is oversampled, the output looks like high frequency digital noise superposed on the input signal. This increases the high frequency noise of the quantized signal, hence lowers the low frequency noise, that is, the noise remaining in the baseband.

The loop filter of Fig. 7.6 plays a dual role. As in other feedback loops it provides gain while it determines the bandwidth of the noise shaper. At low

142 | Delta–Sigma converters

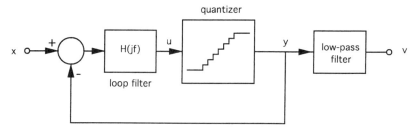

Fig. 7.6 Noise shaping is achieved by means of a non-linear feedback loop.

frequency, the gain is high. The filter compensates the quantization noise almost perfectly. Nothing like that happens at high frequency, however, since the loop gain is cancelled; therefore the noise builds up out of the baseband.

This description shows that low-pass filtering the quantized signal is obviously another essential function to gather the benefits of oversampling. This is done using the filter shown on the right in Fig. 7.6, which ideally cuts everything above f_o and passes everything below. Therefore a high order low-pass filter is needed, which would be very costly if implemented in the analog domain. However, this is not the case because the noise shaper delivers coded data; filtering takes place in the digital domain thus where sharp cut-off characteristics are less of a problem.

The combination of the noise shaper and low-pass digital filter proves that Delta–Sigma converters take the best of analog and digital worlds. The noise shaper is nothing but a fast, relatively inaccurate analog network while the filter is a high precision digital signal processor. This is what makes Delta–Sigma converters good candidates for mixed analog digital circuits, in a world that is dominated by digital technologies.

7.4 A linear approximation of A to D Delta–Sigma converters

In the model shown in Fig. 7.7, the quantizer has been replaced by an independent random noise source e_{rms}. This is nothing but a crude linear approximation of the quantizer noise, since the injected noise is totally independent of the input signal. One ignores the correlation between input signal and quantization noise, as well as the eventual saturation occurring when the signal applied to the quantizer overrules its dynamic range. In spite of these shortcomings, the *SNR*s derived from the model are in good agreement with experimental data. The principal merit of the model is that it helps to understand quantitatively how the performance of Delta–Sigma noise

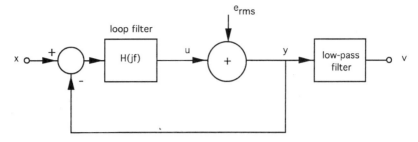

Fig. 7.7 In the linear model of noise shapers, the quantization noise is simulated by means of an independent noise source.

shapers are controlled by the *OSR* and filter order. Predictions based on the same model prove to be satisfactory, as long as the input signal is not more than 10 to 50% of the F.S. depending on the number of bits of the quantizer.

In the circuit of Fig. 7.7, the relation between output y, input x and noise source e_{rms} is determined as:

$$y = H \cdot (x - y) + e_{\text{rms}} \quad (7.10)$$

or rewritten:

$$y = H_x \cdot x + H_n \cdot e_{\text{rms}} \quad (7.11)$$

where:

$$H_x = \frac{H}{1 + H} \quad (7.12)$$

and:

$$H_n = \frac{1}{1 + H} \quad (7.13)$$

which are the transfer functions of the input signal and noise, respectively. These tend to equal one and zero respectively at low frequency, for the filter gain in the baseband is very large. Hence, the output replicates the input and the amount of noise power remaining in the baseband is lowered. It may be evaluated according to the following equation:

$$e^2 = \int_0^{f_o} S(e_{\text{rms}}^2) \cdot | H_n(jf) |^2 \, df = e_{\text{rms}}^2 \frac{2}{f_s} \int_0^{f_o} | H_n(jf) |^2 \, df \quad (7.14)$$

The high-pass transfer function of the quantization noise H_n is actually the starting point of any loop filter synthesis procedure. Once H_n is defined, H can be derived from expression (7.13). Consider for instance the first order high-pass filter H_n below:

$$H_n = \frac{jf}{f_c + jf} \tag{7.15}$$

which according to (7.13) defines an integrator loop filter:

$$H = \frac{f_c}{jf} \tag{7.16}$$

The transition frequency f_c of the integrator must satisfy two goals: it should be smaller than half the sampling frequency to relegate most of the quantization noise outside the baseband, and it should encompass the baseband over a region wide enough in order to provide gain where needed. The larger the gain, the less quantization noise in the baseband. A way to increase the gain without compromising the high frequency quantization noise is to augment the order of the integrator.

Consider the nth order integrator:

$$H = \left(\frac{f_c}{jf}\right)^n \tag{7.17}$$

where f_c equals:

$$f_c = \frac{f_s}{2\pi} \tag{7.18}$$

Note that the factor 2π that divides the sampling frequency is not unique. It was chosen because the transition frequency f_c it defines is the same as that of the discrete integrators considered later when the integrators will be replaced by accumulators to perform non-linear simulations.

Since the baseband gain is large, the noise transfer function H_n in the baseband may be approximated by:

$$|H_n| = \frac{1}{|H|} = \left(\frac{f}{\frac{f_s}{2\pi}}\right)^n \tag{7.19}$$

Thus, the noise power in the baseband according to (7.14) is:

$$e^2 = \frac{\Delta^2}{12}\frac{2}{f_s}\int_0^{f_0}\left(\frac{f}{\frac{f_s}{2\pi}}\right)^{2n} df = \frac{\Delta^2}{12}\frac{\pi^{2n}}{2n+1}\left(\frac{f_0}{\frac{f_s}{2}}\right)^{2n+1} = \frac{\Delta^2}{12}\frac{\pi^{2n}}{2n+1}(OSR)^{-(2n+1)}$$

$$\tag{7.20}$$

A linear approximation of A to D Delta–Sigma converters | 145

and the noise voltage becomes:

$$e = \frac{\Delta}{\sqrt{12}} \frac{\pi^n}{\sqrt{2n+1}} (OSR)^{-\left(\frac{2n+1}{2}\right)} \qquad (7.21)$$

The above expression shows clearly the roles of the noise shaper and quantizer. The contribution of the quantizer is given by the left factor, which is the same as in expression (7.3). The contribution of the noise shaper is shown by the remainder, which offers a sensible way to estimate the combined effects of the *OSR* and filter order '*n*'. This is illustrated by the plot in Fig. 7.8, predicting the potential noise reduction that can be achieved versus the *OSR* and order of several noise-shaping filters. The −10 dB/decade slow rate curve listed under *n* equals zero characterizes the situation where oversampling is taking place without noise shaping (when *n* equals zero, expression (7.20) is identical to (7.8)). Once the noise shaper is introduced, the slope increases by −20 dB/decade each time the order of the loop filter is increased by one unit. As shown, excellent *SNR*s may be reached, leading to the conclusion that high-resolution converters work even with very poor quantizers. A single-bit quantizer, which is the worst situation as to quantization noise, suffices for moderate *OSR*s already if *n* is larger than two. Thus, high order filters are very

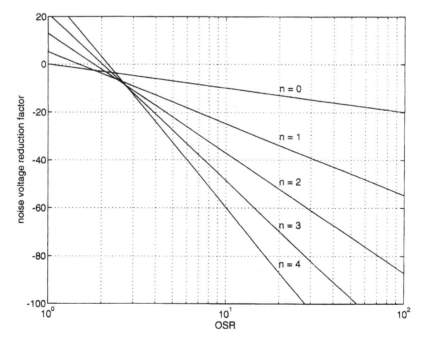

Fig. 7.8 Illustration of the reduction of quantization noise that can be obtained by combining oversampling and noise shaping.

attractive since they yield minimal baseband noise with ever smaller oversampling rates. However, high filter orders are likely to jeopardize the loop stability. They make the design of the loop filter more intricate: as we will see soon, the non-linear character of the quantizer, so far ignored, does not ease the problem.

7.5 The generic Delta–Sigma A to D converter

The architecture of Delta–Sigma A to D converters is that shown in Fig. 7.9. The quantizer consists of two concatenated converters: an A to D followed by a D to A converter. The signal fed back to the input is analog while the data entering the output filter, or decimator, are digital.

Like in any other feedback loop, the reverse feedback path controls the overall accuracy. Consequently the D to A converter must comply with the alleged resolution of the entire Delta–Sigma converter, whereas relaxed tolerances apply to the A to D converter since the latter belongs to the forward path. A simple way to overcome the accuracy problem of the D to A converter is to use a single bit, rather than a multi-bit, quantizer. A single bit quantizer is nothing but a comparator (the A to D converter) followed by a circuit that adds or subtracts a constant reference to the input (the D to A converter). The linearity of the D to A converter is no longer of any importance for its output consists of single steps. Two-bit converters display four steps: the unequal heights of which may impair the accuracy. Nothing like that is possible in a single-bit converter. The step height controls the gain, not the linearity. Of course a large amount of quantization noise will be produced, but a proper choice of *OSR* and filter order takes care of the *SNR*. The graph of Fig. 7.8 clearly shows that

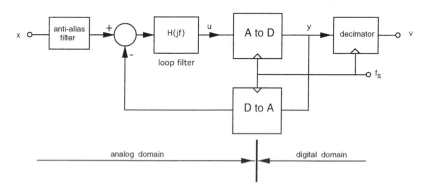

Fig. 7.9 In Delta–Sigma A to D converters, the quantizer is implemented by cascading an A to D and a D to A converter.

Simple first order implementation of a Delta–Sigma A to D converter | 147

a *SNR* improvement as large as 100 dB can be achieved with rather conservative *OSR*s, provided a high order loop-filter is used. For instance, consider a second order Delta–Sigma converter using a single-bit quantizer and requiring a baseband of 8 kHz, like in telephony. According to eqn (7.21), 67 dB improvement is already achieved when the input signal is oversampled by a factor of 30, which corresponds to a sampling rate of nearly 100 kS/s that is easily implemented in CMOS.

Now consider the low-pass filter following the noise shaper. Its purpose is not only to separate the signal from the high frequency quantization noise but also to down-sample the signal to recover the Nyquist sampling rate. Such a filter that turns poorly quantized noise shaped data into high resolution output words is a *decimator*. Its operation is described in Section 7.12.

Note that both the over- and down-sampling operations are transparent because they take place intra-chip. Only the power consumption is enhanced as the result of the oversampling. The increased bandwidth is not a problem.

7.6 Simple first order implementation of a Delta–Sigma A to D converter

The converter shown in Fig. 7.10 is implemented in a digital voltmeter, and is the simplest delta–sigma converter imagined (van de Plassche 1978). It is actually a current-mode implementation of a single-bit first order converter. The noise-shaper consists of a capacitor, a comparator and a clock driven D flip-flop, while the decimator is a bi-directional counter.

The circuit runs more or less like a triangular wave-shaping network. Its functioning is illustrated in Fig. 7.11. The core of the modulator is the capacitor C, which is charged and discharged periodically by two equal and opposite current sources shown in Fig. 7.10. The input current I_{in} introduces a small debalance that slightly tilts the triangular voltage waveform across the capacitor produced by the two current sources. The voltage across the capacitor is checked against the comparator reference (0 V in the example) whose output feeds the D input of a flip-flop. The state of the latter is set by the clock (in the graph, the clock is symbolized by equally spaced vertical dashed lines). When the flip-flop toggles, the sign of the slope across C is inverted.

The output of the comparator in the lower part of Fig. 7.11 shows that situations occur from time to time where the state of the flip-flop does not change in the time interval between one or several clocks. This is the consequence of the debalance produced by the input current. The occurrence of zeros and ones (or minus and plus ones) is an image of the input signal. The difference is nothing but a linear measure of the relative magnitude of the

148 | Delta–Sigma converters

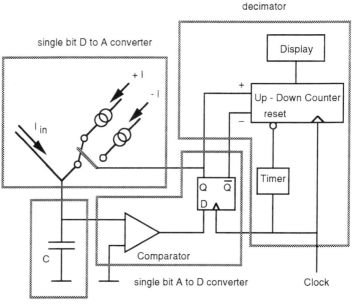

Fig. 7.10 Block diagram of the Delta–Sigma converter described in (van de Plassche 1978).

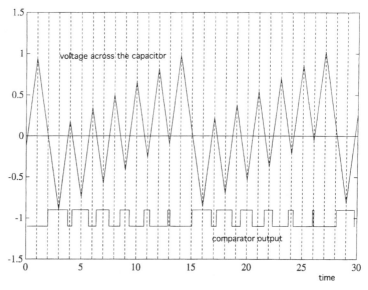

Fig. 7.11 Representation of the capacitor voltage and comparator output of the converter shown in Fig. 7.10.

Simple first order implementation of a Delta–Sigma A to D converter | 149

input current I_{in} with respect to the reference currents. This difference can be evaluated after completing long counting sequences of the bi-directional counter shown on the right in Fig. 7.10. The more time spent, the more accurate the final outcome will be. Thus, this counter not only fulfils the role of a low-pass filter, it also down-samples the data delivered by the flip-flop and increases the length of the output words.

All the ingredients belonging to a true first order Delta–Sigma converter are easily recognizable. The capacitor implements a first order integrator. The comparator is a single-bit A to D converter; the clock driven flip-flop controlling the sign of equal and opposite current sources is a D to A single-bit converter. The counter is nothing but the decimator; and the *OSR* is simply the ratio of the input signal sampling rate over the read-out rate.

The hardware implementation shown in Fig. 7.12 is similar to well-known triangular wave-shaping networks. The integrator is the floating capacitor C located between the two vertical branches in the middle. Current from the upper current source flows left or right according to the state of the flip-flop controlling the ECL-like current switch above. The equal currents delivered by the sources below balance the upper current exactly, thanks to an auxiliary feedback network called 'averaging' network, which maintains the average voltage across the capacitor terminals within the dynamic range

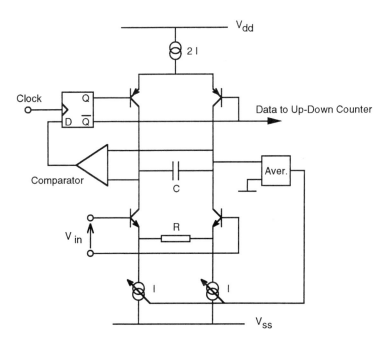

Fig. 7.12 The practical implementation of the converter shown in Fig. 7.10.

defined by the upper and lower current sources. Depending on the state of the flip-flop, current flows into the capacitor from left to right or vice-versa. The degenerated differential pair below translates the input voltage into a current I_{in}, that unbalances the current symmetry. The resulting sawtooth-like voltage across C is applied to the differential comparator driving the D flip-flop.

Although this circuit may be considered as a type of stochastic width modulator, it is a true single-bit first order Delta–Sigma converter. The counter output drives the numerical display of a 4.5-digit voltmeter. Because more than two readings per second is hardly achievable and the input is sampled at two hundred kHz, the oversampling rate is very large. This accounts for the high accuracy, despite the large quantization noise generated by the single bit quantizer and the poor performances of the first-order noise filter.

7.7 More detailed analysis of Delta–Sigma converters operation

The correlation between the input signal and quantization noise does not ease the study of Delta–Sigma converters. Only first and second order converters are amenable to accurate treatment, but the theory is restricted to DC and pure sine waves (Ardalan *et al.* 1987). This section reviews computer simulations

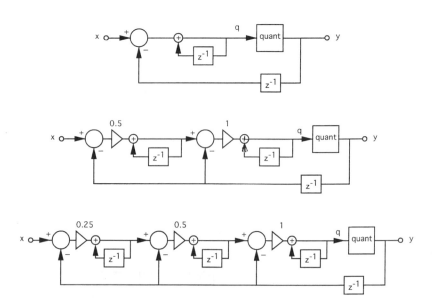

Fig. 7.13 First, second and third order discrete noise shapers.

that give away the independent noise source covered in Section 7.5 (Boser and Wooley 1988).

The three circuits shown in Fig. 7.13 represent discrete implementations of first, second and third order noise shapers. The accumulators are the digital counterparts of analog integrators. The unit-delay closing the feedback loop in each circuit takes care of the clock-cycle delay associated with the quantizer. Other features like distributed feedback, zero-delay accumulators and interstage gains are the result of compromises not covered here.

To become more familiar with the operation of noise shapers, let us select the third order circuit of Fig. 7.13 and consider a 3-bit quantizer. The four plots displayed in Fig. 7.14 show the output of the three accumulators and quantizer considering a sine wave input sampled at 50 times the Nyquist rate. All accumulators deliver signals that reproduce the input sine wave with superposed high frequency noise that tends to increase as we move along the filter. The explanation is shown in the third and fourth plots, which represent the input and output data of the quantizer. The objective is to produce the many code transitions needed to track the input signal as closely as possible. The noise shaper in fact tries to circumvent the lack of subtlety inherent to the quantizer

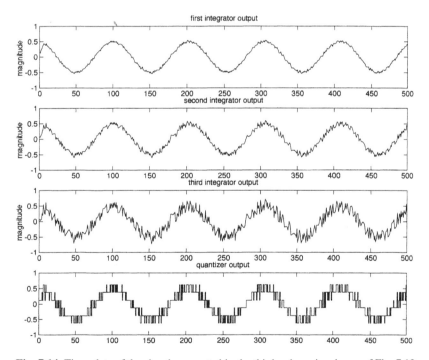

Fig. 7.14 Time plots of the signals generated in the third order noise shaper of Fig. 7.13. The 3-bit quantizer input–output data are illustrated by the two lower graphs.

152 | Delta–Sigma converters

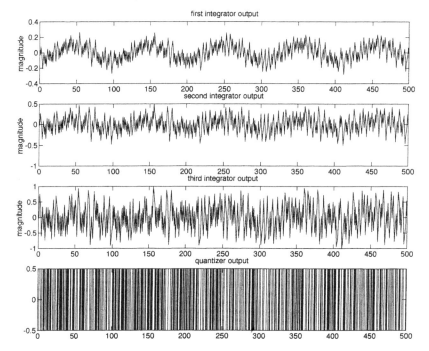

Fig. 7.15 Same as Fig. 7.14 after changing the number of bits of the quantizer from three to one.

output by exploiting the enhanced resolution in time, thanks to the oversampling. This is more apparent even in the single-bit noise shaper considered in Fig. 7.15. Here the output looks like a random square wave. All the data are now exclusively distributed in time since the magnitude has no information. The output looks like an erratic time length modulated binary signal approximating the input signal over time. Note the substantial increase of noise due to the larger step size of the quantizer. The possibility of saturating the integrators and quantizer is larger than in the three-bit quantizer. Therefore the signal represents only 10% of the F.S. instead of 50% as in the three-bit noise shaper.

Looking at the output of the single-bit quantizer, it is hard to believe that this kind of random square wave contains all the data necessary to produce a clean sine wave after low-pass filtering. Nevertheless this is confirmed by Fig. 7.16. The three plots in this figure represent the windowed *fft* transforms of the output signals delivered by the three noise shapers of Fig. 7.13, considering the same input sequence (a sequence of 4096 samples representing 33 entire sine wave periods). The low frequency noise abatement operated at the expense of large high frequency noise is clearly evidenced in the three noise

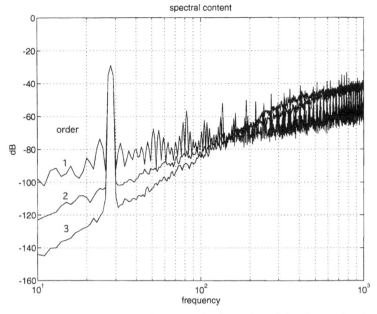

Fig. 7.16 Comparative spectra of the quantization noise of the three noise shapers shown in Fig. 7.13.

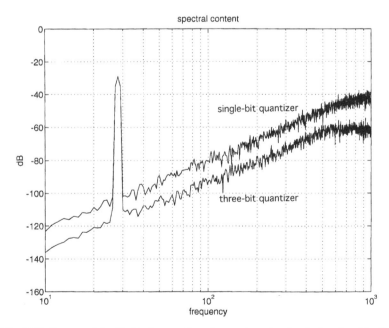

Fig. 7.17 For every bit added to the quantizer, the spectra noise density is reduced by 6 dB.

shapers. Since the power density at low frequency of the noise drops at $-20.n$ dB/decade, where n is the order of the noise shaper, it is obvious that a larger order noise shaper may substantially lower the noise power in the baseband.

According to expression (7.21), every bit added to the quantizer improves the *SNR* by 6 dB. This is ascertained by the plot of Fig. 7.17, where the spectral noise power density is -18 dB lower, when the single-bit quantizer of the second order noise shaper considered above is replaced by a three-bit quantizer.

An estimate of achievable *SNR* can be obtained at this point by assimilating the decimator to an ideal low-pass filter. This is done by integrating the noise power spectral density from zero to f_o to evaluate the total noise power in the baseband. Since an analytic expression of the noise power density in the baseband is needed, a fictitious high-pass filter is introduced. For instance, in the first order noise shaper the noise power density is modeled by the high pass transfer function:

$$|H_n|^2 = k \left(\frac{f}{\frac{f_s}{2}} \right)^2 \qquad (7.22)$$

matching the noise power spectrum. After identifying the factor k, the noise power is derived from the equation:

$$e^2 = \frac{\Delta^2}{12} \frac{2}{f_s} k \int_0^{f_o} \left(\frac{f}{\frac{f_s}{2}} \right)^2 df = \frac{\Delta^2}{12} \frac{k}{5} \left(\frac{f_o}{\frac{f_s}{2}} \right)^5 \qquad (7.23)$$

At this point, the *SNR* is computed and the actual resolution may be derived.

Another method to estimate the resolution is to draw a plot of the *SNR* versus the magnitude of the input signal and find the *ENOB* according to the method described in Section 1.5.1. The procedure illustrated in Fig. 7.18 relates to the third order noise shaper of Fig. 7.13, considering a single-bit quantizer and an ideal low-pass decimator with a cut-off frequency representing 2.4% of the halved sampling rate. The *ENOB* derived from eqn (1.3) reaches 13.3 bits.

This is certainly too optimistic a result for noise power estimations are more demanding, as seen in Section 7.12 when real decimators are contemplated. Most decimators consist of several digital filters in cascade. As we go from one to the next, the sampling frequency is lowered until the Nyquist frequency is reached. Since each filter is a sampled data system in itself, noise is folded back in the baseband and increases the noise power. Thus, a valid evaluation of the noise power implies taking into account the decimator characteristics.

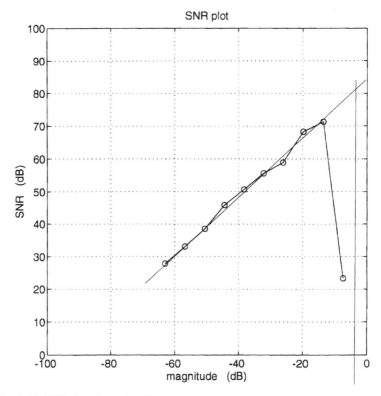

Fig. 7.18 *SNR* plot of a noise shaper.

7.8 Non-linear aspects of Delta–Sigma converters

So far, we have tried to establish similarities between the simple linear model and real noise shapers. The linear approach becomes much more questionable when stability is considered. Noise shapers are circuits that normally oscillate around an equilibrium position. The feedback loop never stops adding or subtracting increments to track the input as closely as possible. Thus, a stable converter is characterized by the fact that the loop-filter produces oscillations that remain bound regardless of the magnitude of the input. This becomes more hazardous when the input nears the F.S. of the quantizer, since saturation enhances the non-linear nature of the noise shaper. Once the quantizer saturates, the signal fed back to the input is stuck and the 'gain' of the feedback loop degrades; the quantizer may need more time to track the input signal again. Eventually the loop could become unstable.

156 | Delta–Sigma converters

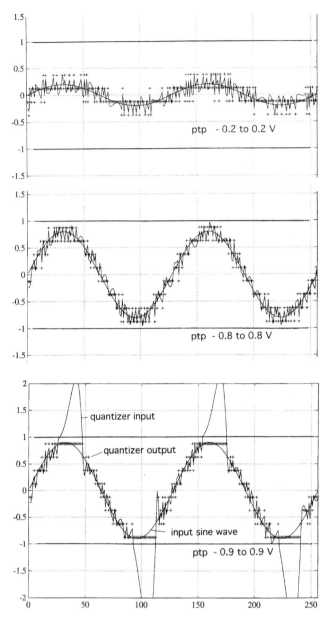

Fig. 7.19 Representation of the input signal (plain line sine wave), output of the second integrator (plain line noise) and quantized output (+) of the second order three-bit noise shaper of Fig. 7.13. Several magnitudes of the input signal are considered: 20% of the F.S. above, 80% middle 90% below. The noise shaper does not operate properly anymore in the last case.

7.8.1 The Limit cycle

Figure 7.19 relates to the second order noise shaper of Fig. 7.13 when considering a three-bit quantizer. The three plots shown in this figure display the input sine wave and the quantizer input and output, considering input signals with increasing magnitudes. When the peak to peak excursion is bound to plus and minus 0.2 V, the quantizer stays within its dynamic range. The is true also when the input is bound between minus and plus 0.8 V. Above, 0.9 V, the noise shaper seems to loose control over the feedback loop because the quantizer output no longer adjusts to the large input signal. Under these circumstances a lot of time is needed to desaturate the quantizer. When fewer quantization levels are considered, the situation may be worse, as shown by the plots of Fig. 7.20. The change here is the single bit quantizer. Its output saturates when the peak-to-peak range exceeds 0.4 V. Instead of a small limit-cycle, the noise shaper locks into a large limit-cycle. The resulting long sequence of identical code words output by the quantizer enhances the harmonic distortion, thus abating the *SNR* and degrading the resolution. This explains the leveling off shown generally by *SNR* plots near the F.S.

The two noise shapers considered so far are stable devices for they always return to normal operation. When the order exceeds two, this does not necessarily hold true. The next section covers the complex problem of stability of high order noise shapers.

7.8.2 Stability of noise shapers

The stability of high order Delta–Sigma converters is still an unsolved problem. Some attempts to find criteria applicable to noise shapers have been reported (Stikvoort 1988; Baird and Fiez 1996). The work of Baird is based on Kalmann's theory of non-linear devices, which assimilates these to linear devices with a variable gain. Kalmann's theory was elaborated to determine stability criteria applicable to weakly non-linear systems like saturated amplifiers. The approach proposed by Baird and Fiez (1994) is similar in that it assimilates the quantizer to a linear amplifier with a variable gain that is equal to the output over input large signal ratio. This gain decays as the magnitude of the input signal increases before collapsing far in saturation. Since the noise shaper is assimilated to a linear network, its behavior may be described in terms of poles and zeros, but unlike linear systems, singularities are not fixed but rather move in the complex plane as the gain changes.

Baird's approach applied root-locus techniques to the signal transfer function $H_x(z)$ to assess the stability of noise shapers. The transfer function has generally one real negative pole outside the unit circle and more poles inside. When the signal applied to the quantizer is small, the 'gain' of the quantizer is large since its output is controlled by the height of the first step,

158 | Delta–Sigma converters

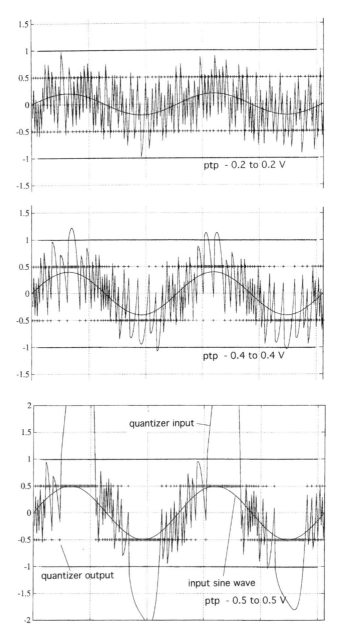

Fig. 7.20 When a single-bit quantizer is used instead of the three bit considered in Fig. 7.19, the noise shaper does not operate correctly anymore once the input signal reaches 50% of the F.S.

Non-linear aspects of Delta–Sigma converters | 159

which may be much larger than the input signal itself. Thus the feedback loop may be potentially unstable. If this is the case, the signal 'u' at the quantizer input (see Fig. 7.9) increases, but while this happens, the quantizer gain drops concurrently. The trend continues until the outside pole has moved inside the unit-circle. The noise shaper then recovers stability, so the signal 'u' starts to decay causing the pole to move out of the circle again. Consequently, the gain increases and the same scenario is repeated. This leads to a simple

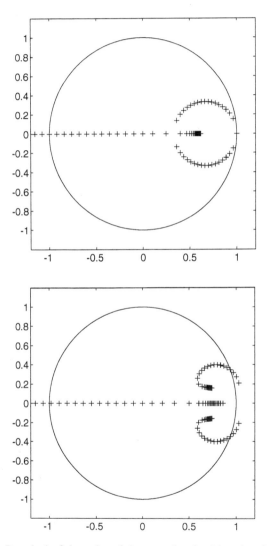

Fig. 7.21 Root-loci of the poles of the second order (above) and third order (below) noise shapers.

interpretation. The pole moving in and out of the unit circle illustrates the oscillatory behavior or the limit cycle tracking the input signal. The oscillation does not imply instability. An unstable noise shaper is traced when one or several other poles move outside the unit circle while the gain decreases. Once outside, these singularities move even further away. The signals get larger and the circuit does not return to normal operation.

Figure 7.21 shows the poles of the $H_x(z)$ function of second and third order noise shapers like those in Fig. 7.13. When the 'gain' of the quantizer, 's', is larger than 1.7, both quantizers have one negative real pole outside the unit-circle; the other poles lie inside. In the second order noise shaper (above), a second pole lies on the positive real axis inside the circle. When the gain 's' decreases, the real negative pole enters the unit-circle while the other pole moves in the opposite direction. After merging, a complex conjugate pair is created whose trajectory corresponds to a circle that ends exactly on the unit-circle. Since no pole other than the first gets out of the circle, the second order noise shaper may be considered a stable device by essence. In the third order noise shaper (below), a different situation is encountered. When 's' gets smaller than some critical gain (0.4 in the example), the pair of complex conjugate poles moves out the unit-circle. As the gain gets smaller, these move further outside the circle. Then nothing prevents the noise shaper from drifting away indefinitely until it gets locked. If the input is bound, however, so that the limit-cycle never reaches the critical gain, the noise shaper will operate safely. Hence, whether a high order noise shaper is stable or not is a difficult question. If for some reason the signal gets large, a safe noise shaper may become eventually unstable. The question then arises as to how to correct such a poor situation. A possible strategy is to reset all the accumulators. The noise shaper then operates correctly again hoping that a similar situation does not reoccur. Of course, if this happens too often, the SNR and resolution will be adversely affected. A better approach is to widen the dynamic range of the quantizer, for instance consider a two-bit quantizer. This reduces the chance of running into large limit-cycles without jeopardizing the accuracy, since the impairments of the multi-bit D to A converter do not relate to normal operation.

Interestingly, the root locus method is a simple means to show the superiority of the stability of multi-bit over single-bit quantizers. In single-bit quantizers, the gain near zero is very large. In multi-bit quantizers, it remains close to one. Once the quantizer saturates, the gain starts to decay like in the single-bit quantizer. Larger input signals may be tolerated before the noise shaper becomes unstable. This is similar to the conclusions drawn from the plots shown in Figs 7.19 and 7.20.

7.8.3 Idle tones

Another typical feature of noise shapers is the emergence of *idle tones*. This is the appearance of quasi-periodic sequences, which deteriorate the *SNR* through the introduction of spurious rays in the baseband, as in the first order noise shaper shown in Fig. 7.16. To understand this phenomenon, let us reconsider Fig. 7.11, which relates to the delta–sigma digital voltmeter described in Section 7.6. The triangular waveform portrayed in the figure illustrates the data input to the quantizer of the first order noise shaper. It is obvious that after a number of clock pulses the same pattern occurs more or less with slight differences. If this situation repeats itself for a certain time, sub-harmonics are likely to be created.

However, this strongly depends on the input signal. A little change suffices to break the periodicity. Of course, large OSRs help to counteract idle tones also, keeping their contributions, outside the baseband. A more consistent way to deal with the problem is to add dither noise (Magrath and Sandler 1995) or chaos (Schreier 1994). A little amount of randomness suffices to reduce the amount of idle tones and make the task to build new ones more difficult. Of course, the dither noise deteriorates the *SNR* somewhat. A better solution is to make use of higher order loop-filters, since the large amount of processing inside the loop filter tends to fuzzify the signals even more and render dither noise useless. The spectra of the second and third order noise shapers of Fig. 7.16 ignore spurious peaks and confirm the above statement.

7.9 Discrete versus continuous loop filter implementations

Essentially loop-filters are analog filters. There are two possible implementations: switched capacitor or continuous filters. They differ as far as power consumption, sensitivity to clock jitter and contribution to aliasing.

In the continuous filter approach, sampling takes place in the quantizer after filtering. In the switched capacitor approach, the filter samples the difference between input data and quantizer output. Hence, the Op Amp bandwidth should be at least five times larger than the clock frequency to prevent the accumulation of errors from incomplete charge transfer that abate the *SNR*. Continuous filters do not suffer from this. Therefore, their bandwidth can be smaller so they require less power. However, continuous filter implementations are more sensitive to clock jitter.

Another peculiarity relates to the anti-aliasing strategies. As cited in Chapter 4, the part of the input signal spectrum that exceeds half the sampling frequency folds back in the converter baseband. To avoid this, an analog anti-alias filter is located in front of the converter. The specifications of this filter are fixed by the cut-off frequency of the baseband, the sampling

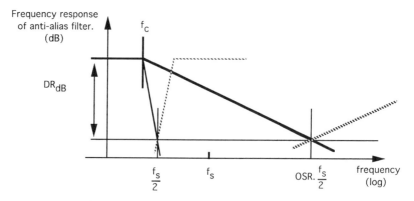

Fig. 7.22 Oversampling results in relaxed anti-alias specifications.

frequency, and the magnitude or dynamic range DR_{dB} over which spectral lobes should not overlap. When the signal is sampled at the Nyquist rate, the cut-off frequency and the halved sampling frequency may get so close that a high order low-pass filter is necessary to guarantee the separation. This anti-alias filter is very costly for it must be implemented in the analog domain. The anti-alias requirements of Delta–Sigma converters are substantially relaxed as a result of the oversampling as shown in Fig. 7.22. The order of the anti-alias filter is given by the expression:

$$n = \frac{DR_{dB}}{20 \log_{10}\left(\frac{f_N}{f_0}\right) + 20 \log_{10}(OSR)} \quad (7.24)$$

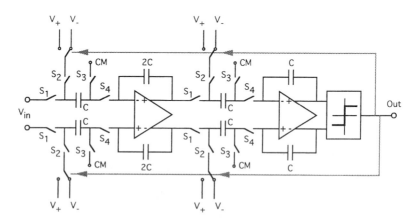

Fig. 7.23 A switched capacitor implementation of a second order noise shaper.

Let us compare a Nyquist-rated and an oversampled converter assuming that the anti-alias specifications impose 60 dB attenuation at the point where the spectral contribution of adjacent lobes cross one another. According to expression (4.4), a Nyquist-rated sampler requires a tenth order filter to keep the cut-off frequency of the input signal one octave below half the sampling frequency. In the Delta–Sigma converter, a third order filter is sufficient considering an *OSR* of 20. A simple RC or RLC network does the job. Moreover, if the loop filter is a continuous filter, the anti-aliasing function is taken by the loop filter since sampling takes place in the quantizer.

The circuit of Fig. 7.23 shows a typical switched capacitor implementation of the second order noise shaper of Fig. 7.13. The filter consists of two cascaded fully differential integrators. Often the same gain is chosen for both integrators.

Current mode versions of the loop filter have also been investigated (Daubert and Vallancount 1991). Dynamic current memories are good candidates, especially for low-voltage converters. In Moeneclay and Kaiser (1997), a 14-bit current-mode Delta–Sigma converter is described. A recent low-voltage architecture that uses switched Op Amps operating under 0.90 V with a dynamic range of 77 dB is described in Peluso *et al.* (1998).

7.10 Widening the bandwidth

The abatement of quantization noise by means of large oversampling rates is hindered by the fact that sampling frequencies cannot increase indefinitely, since they are bound by the technology. When the bandwidth gets too large to conform to acceptable sampling rates, the *OSR* must decrease. To counteract the resulting loss of resolution, one may either increase the order of the loop filter or contemplate other means, like multi-bit quantization, cascaded loop quantizers and interpolative modulators.

7.10.1 Synthesis of high order noise shapers

The design procedures of high order noise shapers rely on the assumption that they can be modeled adequately by means of linear feedback loops despite the non-linear behavior of the quantizer. The noise shaper is assimilated to the linear model in Section 7.4 so that the design of the loop filter is greatly simplified because well known synthesis procedures may be used. However when the quantizer enters saturation its actual behavior is hard to predict. Computer simulations are the only way to answer the question.

The z transform of the linearized first order noise shaper shown in Fig. 7.13 is given by the equation:

$$Y(z) = z^{-1}X(z) + (1 - z^{-1})E(z)$$
hence $H_x(z) = z^{-1}$ (7.25)
and $H_n(z) = (1 - z^{-1})$

X, Y and E are the z transforms of the input x, the output y and the noise source e_{rms}. The noise transfer function has one zero located at plus one. The same result holds true for the higher order filters of Fig. 7.13. Their noise transfer functions exhibit multiple zeros at the same position according to the equation:

$$H_n(z) = (1 - z^{-1})^n \qquad (7.26)$$

where n is the filter order.

These noise shapers are far from optimal for their noise at high frequency tends to increase concurrently with the order filter, as shown by the high frequency noise of the plots in Fig. 7.16. Large amounts of high frequency noise favor the creation of unwanted limit cycles and endanger stability near saturation. The synthesis of high order noise shaping filters is therefore not only a matter of low frequency specifications but also of high frequency noise tolerances.

Synthesis starts from the required noise transfer function H_n. The loop filter transfer function H is derived from H_n as outlined in Section 7.4. The 4th

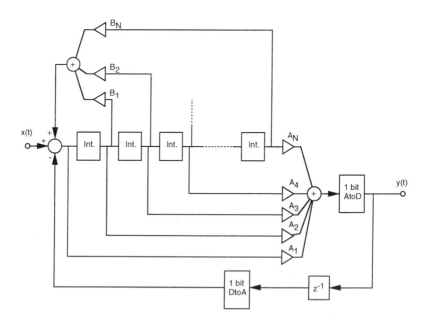

Fig. 7.24 The loop-filter proposed in (Chao et al. 1990).

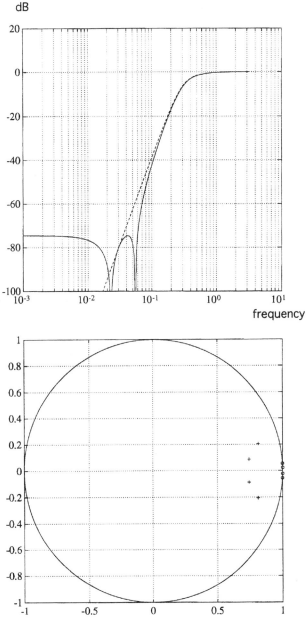

Fig. 7.25 (Above) Noise transfer characteristics of the noise shaper described in (Chao *et al.* 1990): the basic fourth order Butterworth characteristic (dashed lines), the same after moving the zeros along the unit-circle (plain lines). Below the poles and zeros in the second version.

order noise shaper described in Chao *et al.* (1990) is a good example. The signal and noise transfer functions are respectively:

$$H_x = \frac{H}{1 + z^{-1}H} \qquad (7.27)$$

and

$$H_n = \frac{1}{1 + z^{-1}H} \qquad (7.28)$$

The loop filter is a cascade of integrators that transfer weighted data in both forward (A coefficients) and reverse (B coefficients) directions. The noise transfer function H_n derived from (7.28) of the circuit shown in Fig. 7.24 is given as:

$$H_n = \frac{(z-1)^N - \sum_{i=1}^{N} B_i(z-1)^{N-i}}{z\left[(z-1)^N - \sum_{i=1}^{N} B_i(z-1)^{N-i}\right] + \sum_{i=0}^{N} A_i(z-1)^{N-i}} \qquad (7.29)$$

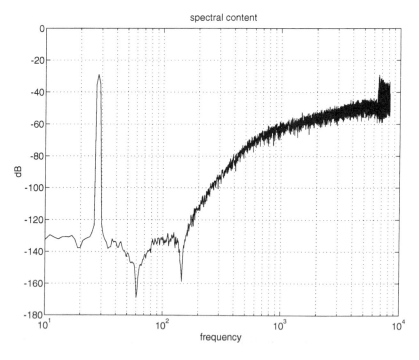

Fig. 7.26 The quantization noise spectral density (16384 points *fft*) of the noise shaper described in (Chao *et al.* 1990).

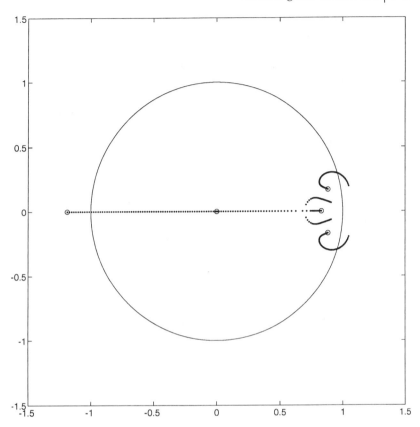

Fig. 7.27 Root-loci of the poles of the noise shaper described in (Chao *et al.* 1990).

The problem is to determine which loop filter coefficients A_i and B_i yield the best noise transfer function H_n. These are determined in two steps. First, all the B_i's are assumed zero and the A_i's are chosen to implement a fourth order high-pass Butterworth filter. The filter suppresses most of the noise in the baseband and prevents the high frequency noise from surpassing limits. The target transfer function in the z-plane is derived from the bilinear transform of the 4th order filter in the s-plane. Its characteristic is shown as the dashed curve in the upper part of Fig. 7.25. All A_i coefficients are determined by solving a set of linear relations identifying the denominators of the z-plane target transfer function and of the above expression. The lower plot in Fig. 7.25 shows the locations of the four poles. The noise transfer function H_n has four zeros at the point where z equals one. Let us move these zeros from their original position along the unit circle to introduce two zeros in the baseband. Consequently the dashed transfer curve of Fig. 7.25 becomes the plain curve.

The much steeper high-pass characteristic obtained near the end of the baseband slightly widens the bandwidth. But more noise is experienced in the baseband as a result of the two zeros. As long as this noise does not compromise the converter resolution, the widened baseband is an improvement. To optimize the positions of the zeros, the corresponding B_i coefficients are chosen to fit a Chebyshev polynomial approximation.

The 16384 point FFT plot of Fig. 7.26 represents the spectral noise power density of the corresponding single-bit noise shaper. The two zeros and the steep increase of quantization noise near the edge of the baseband are clearly visible. The noise power density inside the baseband is approximately −130 dB.

The root locus plot of Fig. 7.27 gives an idea of the behavior of the above noise shaper near saturation. Since the denominators of H_x and H_n are the same, this root locus is derived easily from expression (7.29). All that is necessary is to multiply the coefficients A_i by the variable gain 's' of the quantizer. There are five poles: one at zero, two on the real axis and two forming a pair of complex conjugate poles. The pole at zero is the results of the unit-delay in the return path that takes into account the delay associated with the quantizer. It is the only pole that does not move with the gain. The pole remaining on the real axis outside the unit circle is the one that fixes the limit cycle; all the other poles are inside the unit-circle. Lowering the gain nears the two real poles to one another until they form another pair of complex conjugate poles. The other pair moves outside the unit-circle when 's' equals 0.40, which is the critical 'gain' beyond which the noise shaper becomes unstable.

The sensitivity to circuit imperfections of the above noise shaper was extensively studied and the results are reported in Chao *et al.* (1990). The authors report that the Op Amp gain must be at least 60 dB and the slew rate equal to 60 V/μs. As far as the filter coefficients, errors between 5 and 30% are tolerated owing to the relative insensitivity of the Butterworth filter to coefficient errors.

A wealth of other loop-filter implementations are reported in the literature. Most topologies are divided in categories: integrators with weighted feed-forward summation and local resonator feedback, chains of integrators with distributed feedback, combinations of distributed feedback and feed-forward inputs. A survey of various topologies can be found in Norsworthy *et al.* (1997).

7.10.2 Multi-bit versus single-bit noise shapers

The compromise of resolution versus oversampling rate, plus order of the noise shaper sets an upper limit on the achievable bandwidth that can be achieved with Delta–Sigma converters. The ultimate sampling rates are set by technology and the order of the noise shaper is bound by stability considerations. Consequently, bandwidths in the MHz range are hard to

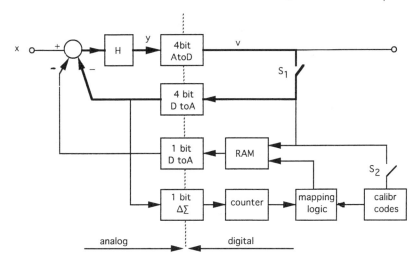

Fig. 7.28 Multi-bit noise shaper with auto-correction.

achieve. The only remaining issue is to reduce the oversampling rate and to compensate the subsequent *SNR* loss by increasing the resolution of the quantizer. Every bit added to the quantizer improves the *SNR* by 6 dB, as stated in expression (7.21). The A to D and D to A converters must be parallel devices so that they do not slow down the sampling rate. The converter in the feedback loop is generally a thermometer coded scaled D to A converter and the A to D converter in the forward branch a flash converter (flash converters are reviewed in Chapter 8). Since the resolution of quantizers rarely exceeds 5 or 6 bits, the number of discretization levels is relatively small. Thus both area and power consumption are affordable.

The accuracy of the flash converter is negligible since it is a part of the forward path in the feedback loop. On the other hand, D to A converter performances play a dominant role as in any feedback loop. Disparities in step sizes distort the signal fed back, and thus increase the harmonic distortion of the noise shaper. In single-bit quantizers, the problem is ignored since the D to A converter only outputs two levels; when more than a single bit is considered, the *DNL* becomes essential.

The problem may be eluded eventually through the combination of a multi-bit A to D and a single-bit charge integration D to A converter (Leslie and Singh 1992). The linearity is preserved but the integration of repeated charge packets cycle after cycle lengthens the overall time, making this a counterproductive solution regarding bandwidth. Trimming, albeit expensive, is a possibility given the small numbers of elements to adjust. But more practical approaches are preferable, such as *auto-correction* and *mismatch randomization*.

Figure 7.28 illustrates the *auto-correction* implementation suggested by Baird and Fiez (1996). The linearity of the multi-bit D to A converter is tested by means of an auxiliary single-bit Delta–Sigma A to D converter operating during idle periods. Errors are corrected on-line when conversions take place. In the auto-calibration mode, switch S1 is open and S2 closed. The calibration code generator issues the entire set of code words that control the N-bit D to A scaled converter. A counter performs decimation like that used in the simple Delta–Sigma voltmeter considered in Section 7.6. The mapping logic block evaluates errors from the multi-bit D to A converter and the corrections are stored in the RAM. Once calibration is completed the states of both switches are exchanged. Data from the main Delta–Sigma converter control the multi-bit D to A converter and drive the RAM that controls the fine D to A correction converter at the same time. The analog output of both the multi-bit and the correction converters are added, closing the feedback loop of the noise shaper.

The *mismatch randomization* technique differs in that instead of correcting the multi-bit D to A converter, harmonic distortion from mismatches is changed into wide band noise. Since this noise spreads over the full Nyquist bandwidth, it contributes less to baseband noise. The idea is to randomize the errors of the thermometer-coded D to A converter by selecting the unit-elements in a pseudo-random manner (Leung and Sutarje 1992). This is a good way to avoid repetitions of identical combinations of unit-elements for the same code words. The ensuing wideband noise is added to the quantization noise floor, because the noise shaper does not differentiate between noise sources. A number of strategies have been proposed that aim to minimize the noise left over in the baseband. In the Data Weighted Averaging (D.W.A.) technique, the unit-elements form a loop. They are selected by means of a pointer. Since the spectrum resulting from unit-elements mismatches is roughly noise shaped, the contribution to the *DNL* and *INL* errors of the D to A converter are substantially reduced. A multi-bit Delta–Sigma converter that uses weighting averaging and reaches 19-bit resolution is described in Nys and Henderson (1996).

7.10.3 Cascaded (MASH) converters

Cascaded noise shapers are also called MASH noise shapers (Multistage Noise Shapers). They offer an interesting alternative for widening bandwidths without increasing the sampling frequency while avoiding the stability problems inherent to high order noise shapers. MASH noise shapers are implemented by cascading several low order, thus unconditionally stable, noise shapers (Uchimura *et al.* 1988). For instance, a second order noise shaper cascaded with a first order is equivalent to an unconditionally stable

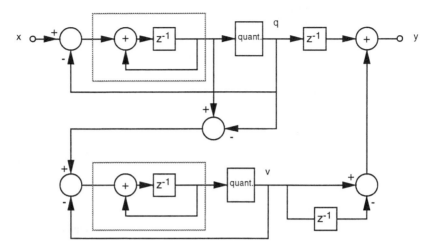

Fig. 7.29 A cascade of two first order noise shapers is equivalent to a second order noise shaper (MASH).

third order noise shaper. More devices can be cascaded to further increase the overall order. However, the various noise shaper's output must be matched to obtain good performances, since MASH converters exploit noise cancellation techniques and are thus sensitive to imbalances among the data delivered by the sub-noise shapers. Excellent results suppose that mismatch sources (Op Amps, gains, etc.) are under control and the number of cascaded noise shapers is not too large.

A simplified second order MASH noise shaper consisting of two cascaded first order noise shapers in shown in Fig. 7.29. The underlying idea is to measure the quantization noise of the first noise shaper with the second and to digitally cancel the noise of the first. The actual quantization noise is evaluated by taking the difference between the input and output data of the first quantizer and applying this difference to the second noise shaper. The filtered noise is then subtracted from the output of the first noise shaper after an appropriate transformation summarized in the following.

We start from the first noise shaper output, which according to expression (7.25) is:

$$Q(z) = z^{-1}X(z) + (1 - z^{-1})E_1(z) \tag{7.30}$$

The noise from the first noise shaper is injected in the second noise shaper once its sign has changed. The output of the second noise shaper is given as:

$$V(z) = -z^{-1}E_1(z) + (1 - z^{-1})E_2(z) \tag{7.31}$$

The second term represents the additional noise created by the second noise shaper. The output of the two noise shapers are combined to cancel the contribution of the first noise shaper. Therefore we differentiate the output of the second nose shaper and add the result to the delayed output of the first noise shaper, resulting in:

$$Y(z) = z^{-1}Q(z) + (1 - z^{-1})V(z) \qquad (7.32)$$

which is equivalent to:

$$Y(z) = z^{-2}X(z) + (1 - z^{-1})^2 E_2(z) \qquad (7.33)$$

Provided exact cancellation of the noise from the first device can be achieved, the noise produced by the MASH converter is the same as that from a single second order noise shaper. In other words, the spectral noise power density drops at the rate of –40 dB per decade. Since noise cancellation occurs in the analog domain, the Op Amps should be identical. Similarly, the analog sum must be performed without introducing additional errors. In practice, the *SNR* improvement is generally substantial, but becomes questionable when more than two or three noise shapers are cascaded.

The cascaded approach is a reliable alternative whenever stability is a major concern. Thus, MASH noise shapers still receive wide consideration and are subject to ongoing research. Benabes *et al.* (1996) introduced a promising multi-stage closed-loop approach known as the MSCL. The combination of multi-bit and cascaded Delta–Sigma algorithms seems to offer very good performances, as illustrated in Brooks *et al.* (1997) which describes a 16-bit Delta–Sigma converter with 2.5 MHz data rate.

7.11 Bandpass A to D Sigma–Delta converters

A new class of noise shapers is obtained when the frequency response of the loop filter is changed from lowpass to bandpass. In these the quantization noise remains outside the passband since the feedback loop is operative only within the bandpass. This is known as a *bandpass* Delta–Sigma converter (Schreier and Sneigrove 1989). The rational behind such converters lies in possible RF applications. In conventional radio systems for instance, the detection of the *I–Q* components is usually achieved by mixing the analog output from the intermediate frequency bandpass filter with two 90° phased components of the local oscillator. After mixing, the signals are band-limited and converted into digital words for further processing. Care must be taken to guarantee the correct discrimination of the output data. The matching tolerances of the mixers are very strict, typically 0.1 dB in magnitude and 1 degree in phase for 40 dB rejection.

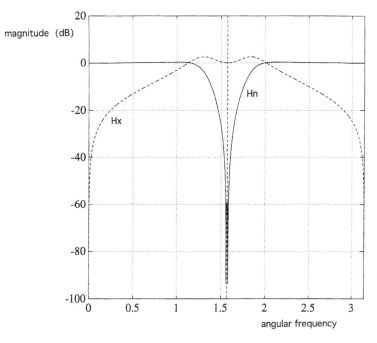

Fig. 7.30 The signal transfer function (dashed lines) and noise transfer function (plain lines) of the 4th order bandpass Delta–Sigma converter described in (Jantzi *et al.* 1993).

Bandpass Delta–Sigma converters ease the implementation because the signal is digitized prior to mixing. Consequently, problems associated with DC offsets and $1/f$ noise are eluded. Thus bandpass single-bit Delta–Sigma converters are definitely an interesting alternative. The local oscillator controlling the mixing only delivers simple digital sequences (0,1,0, −1 for the sine wave and 1,0, −1,0 for the cosine-wave local oscillator).

Bandpass Delta–Sigma converters are synthesized in the same manner as their low-pass counterparts. Replacing z^{-1} by $-z^{-2}$ changes the accumulators used in the noise shapers of Fig. 7.13 into resonators. In terms of poles and zeros, this is the same as splitting the singularities of low-pass noise shapers in two entities that move away from their initial position symmetrically along the unit-circle.

A brief review follows of the 4th order bandpass noise shaper described in Jantzi *et al.* (1993). As usual the starting point is the noise transfer function H_n. The pole-zeros choice has the same constraints as to causality and noise limitation outside the bandpass to reduce the risk of instability. The signal and noise transfer functions shown in Fig. 7.30 are the results of a configuration of

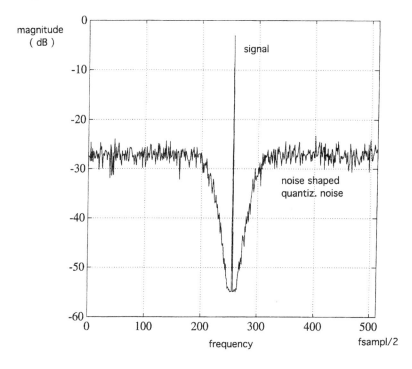

Fig. 7.31 The noise spectral density of the linear model of the bandpass noise shaper.

poles and zeroes, consisting of two pairs of complex poles and two pairs of zeros on the unit-circle near plus and minus $j\pi/2$:

$$\begin{aligned}\text{zeros} &= e^{j(\pm\frac{\pi}{2}\pm 0.0111)}\\ \text{poles} &= \pm 0.2206 \pm j0.7973\end{aligned} \quad (7.34)$$

The noise spectral distribution shown in Fig. 7.31, which was obtained using a linear approximation of the noise shaper shows clearly the input sine wave surrounded by depressed noise in the bandpass region.

Much effort has been made to improve the performances of bandpass Delta–Sigma converters as far as possible, by taking advantage of the submicron CMOS technology. Continuous filter implementations seem very promising for IF frequencies beyond 10 MHz. A number of state of the art contributions are presented in the December 1997 issue of *JSSC* (Jantzi *et al.* 1997) and the May 1998 issue of *IEEE Trans* on Circuits and Systems (Chuang *et al.* 1998), (Bazarjani and Snelgrore 1998).

7.12 The decimator

Decimators are an essential component of Delta–Sigma A to D converters. They attenuate the out of band noise and resample the signal at the Nyquist rate. Since their design is not trivial, a lot of tools have been developed based on computer programs (Leung 1991).

In the first order Delta–Sigma converter considered in Section 7.6, the decimator consisted of a counter or accumulator which was reset periodically after reading out its content. The accumulator's purpose was not only to average out the data from the noise shaper over long sequences, but also to resample the data from the noise shaper to substitute N-bit words to the single-bit data stream entering the decimator. The filter is currently designated as the '*accumulate-and-dump*' first order decimator. It is very straightforward and requires nothing but simple digital circuitry, which is the cornerstone of most decimators.

Since the noise in the baseband after noise shaping is usually very small and rises at a slow rate outside the baseband, a fairly simple filter like the one above suffices to achieve reasonable *SNRs*. Unfortunately, its rapid fall-off at low frequency has an adverse effect on the baseband. Other filters, like F.I.R. and I.I.R. filters, lend themselves to closer control of the baseband transfer characteristic; the trouble with these filters is that they become increasingly expensive with regard to area and power consumption when the *OSR* becomes large. The combination of two filters, an accumulate-and-dump and a F.I.R. filter, offers a better way to achieve a satisfactory frequency response while not requiring too large an area. Therefore, decimators are usually multi-stage devices. A kth-order accumulate-and-dump or sinc^k filter is used to first down-sample the signal from f_s until some intermediate frequency f_d, which may be four or eight times the baseband. Then an F.I.R. filter reshapes the baseband characteristic to keep either the magnitude of the input signal or the delay constant.

The z transform of the first order accumulate-and-dump decimator is given by the expression:

$$D(z) = \frac{1}{M} \cdot \frac{1 - z^{-M}}{1 - z} \qquad (7.35)$$

where M is the number of counts between consecutive reset pulses. Transposed in the frequency domain, the transfer characteristic is given by the expression:

$$D(f) = \frac{\text{sinc}\left(M\frac{f}{f_s}\right)}{\text{sinc}\left(\frac{f}{f_s}\right)} \qquad (7.36)$$

$$\text{where}: \text{sinc}(x) = \frac{\sin(\pi x)}{\pi x}$$

176 | Delta–Sigma converters

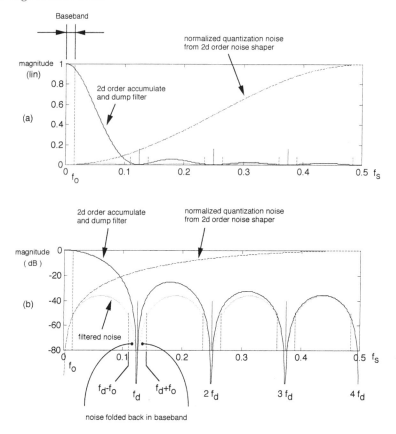

Fig. 7.32 Linear (above) and log (below) scaled frequency response of a second order accumulate-and-dump decimator (plain lines), normalized spectral quantization noise density of a second order noise shaper (dashed lines) and noise density after the accumulate-and-dump filter (dotted lines). The vertical line left represents the baseband. Noise encompassing the zeros of the comb filter within the baseband widths folds back into the baseband.

If the filter order exceeds one, the above expressions become respectively:

$$D(z) = \frac{1}{M^k} \cdot \left(\frac{1 - z^{-M}}{1 - z} \right)^k \tag{7.37}$$

and

$$D(f) = \left(\frac{\mathrm{sinc}\left(M \frac{f}{f_s}\right)}{\mathrm{sinc}\left(\frac{f}{f_s}\right)} \right)^k \tag{7.38}$$

where k is the order of the filter.

Figure 7.32 shows the performances of an ideal second order *sinc* filter followed by a fictitious ideal low-pass filter, which portrays a boxcar F.I.R. filter. The input data, oversampled by a factor of 64, are down-sampled in two steps, a factor of 8 in the *sinc* filter and another factor of 8 in the F.I.R. filter. The sampling frequency, intermediate and baseband frequencies are respectively f_s, f_d and f_o. The plain line represents the frequency response of the accumulate-and-dump filter and the dashed vertical line on the left, the ideal baseband characteristic. For an intuitive idea of the potential noise power abatement, the normalized spectral noise density of a second order noise shaper is plotted in the same graph. It is represented by the dashed surelevated cosine curve that starts on the left at zero and peaks on the right at one. The noise density after the *sinc* filter is represented by the dotted line. Since the transfer characteristic of the *sinc* filter exhibits zeros at every multiple of f_d, the noise density follows the same trend. There are several contributions to the noise power in the baseband after the F.I.R. resampling. First, of course, the noise collected in the baseband, from minuse to plus f_o. Second, the noise foldback from all the bands centered around f_d and multiples, between minus and plus f_o. This noise is sampled at the rate f_d and therefore folded back in the baseband. In the log-scaled plot below, the noise contributions are easily estimated. Luckily the *sinc* filter has zeros at f_d and all its multiples so that the large noise at high frequency does not overrule the decimator performance.

Decimators are always a compromise of several factors. Increasing the down-sampling frequency of the *sinc* filter reduces the noise but has a detrimental effect on the edge of the baseband. The attenuation near the edge gets larger and may be even too large to allow satisfactory compensation by means of the F.I.R. filter for it may be too constraining to implement. A good way to reduce noise in the baseband without increasing the down-sampling rate of the F.I.R. filter is to use multirate filters (Crochiere and Rabiner (1981) or poly-phase IIR filters (Ma 1992). Their frequency response is roughly the same as that of *sinc* filters, but instead of a single zero in the middle of the aliased bands, these filters possess additional zeros uniformly distributed throughout the baseband. The result is a kind of Chebyshev-like stopband characteristics around each multiple of the intermediate frequency. These filters are naturally more complex than the accumulate-and-dump filters. However they are compulsory when high performance noise shapers like that described in Section 7.10.1 are considered (Sherstinsky and Sodini 1990); otherwise the noise near the edges of the baseband annuls the benefits from the zeros of the noise shaper.

The F.I.R. filter consists generally of a digital filter with either a tapped delay line or a shift register to perform the weighted summations. A dedicated signal processor also may be used to perform this step. The sizes of the F.I.R. filter weighting coefficients are generally modest (6 to 8 bits).

7.13 D to A Delta–Sigma converters

This last section shows that the principles underlying A to D Delta–Sigma converters are not restricted to these converters, but are equally appropriate for D to A converters. Oversampling and noise shaping are the key issues as in A to D converters. The input is a sequence of coded words sampled at the Nyquist rate, that must be oversampled to begin with. While this is done

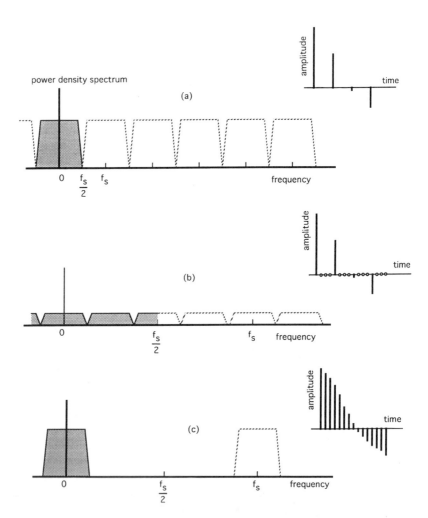

Fig. 7.33 Interpolation increases the sampling rate of digital data. Spectrum of the original Nyquist-sampled data (above), after the introduction of three zeros between each sample (middle), and after low-pass filtering and amplification (below).

D to A Delta–Sigma converters | 179

simply by increasing the sampling frequency in A to D converters, oversampling data from a CD player requires a different procedure since more data cannot be directly accessed. This is done through *interpolation*, the dual of decimation. Decimators deliver multi-bit words at the Nyquist rate, interpolators do the opposite. They oversample the digital data to drive a lower resolution D to A converter whose output data are low-pass filtered by means of a simple analog filter. The operation is the same as in A to D converters, but follows an opposite sequence.

Interpolators add data to the input sequence by inserting zeros between original samples. The operation is performed in a digital filter that acts as a reverse decimator, for instance, three zeros put between every input sample as illustrated in Fig. 7.33. Plot 'a' represents the power density spectrum of the input digital data. As should be expected the baseband encompasses exactly half the sampling rate. In 'b', three zeros are inserted between every input sample so that the actual sampling frequency is multiplied by four. As a result, below half the sampling frequency the spectrum consists of the baseband spectrum plus four side lobes whose magnitudes are divided by four. If the three middle lobes were removed and the remaining spectrum multiplied by four, a spectrum such as 'c' would be obtained. This spectrum characterizes the same signal as 'a' but relates to a fourfold sampling rate. Thus, all that is needed to perform interpolation is a digital low-pass filter that removes unwanted lobes from the sequence 'b' and multiplies the output signal by the *OSR*. Since these operations are done in the digital domain, they are performed with high accuracy.

The algorithm documented by Fig. 7.34 illustrates the interpolation steps of a signal consisting of the sum of two incommensurable sine waves whose magnitudes are respectively 1 and 0.3. The crosses on the plain line below, mark the input data before interpolation. Inserting three zeros between every sample, obtains the upper spectrum. The two rays on the left illustrate the original signals whereas the replicas in the middle and on the right are the results of zeros inserted between the input samples. In plot 'b', the oversampled sequence is filtered by a low-pass Remez filter whose frequency response is shown in the same plot. Once the unwanted side lobes are attenuated, the output signal below, represented by crosses, is obtained. The samples are clearly a fourfold version of the input, slightly delayed in time with respect to the input because of the filter.

Like decimation, interpolation generally occurs in two steps. For instance, the sampling frequency is multiplied first by four, followed by a much larger factor, e.g. 64, during a second step. This is to prevent the steep cut-off characteristic of the filter, needed to separate the baseband from the first sidelobe. This is the most demanding part of the filtering, and is generally done using an F.I.R. filter taking advantage of the low *OSR*. The rest is done by a comb filter just like in A to D Delta–Sigma decimators, but the other way around.

180 | Delta–Sigma converters

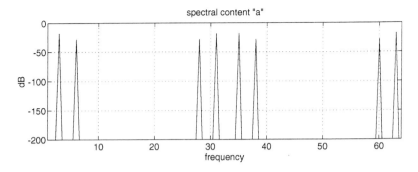

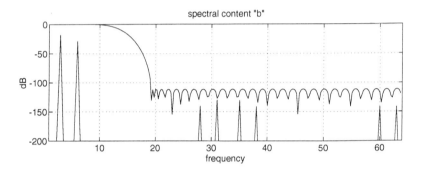

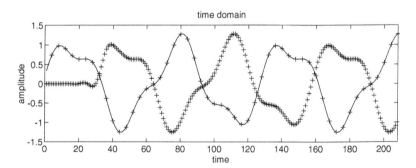

Fig. 7.34 Above, the *fft* of a signal consisting of two superposed sine waves after introduction of three zeros between each sample. Middle, the same after filtering (the frequency response of the filter is shown also). Below, the original sampled signal (plain lines) and interpolated signal (+).

The goal is to replace Nyquist-rated high resolution words with oversampled low resolution data, which track the input as closely as possible, so that the number of bits of the output D to A converter can be much smaller.

D to A Delta–Sigma converters | 181

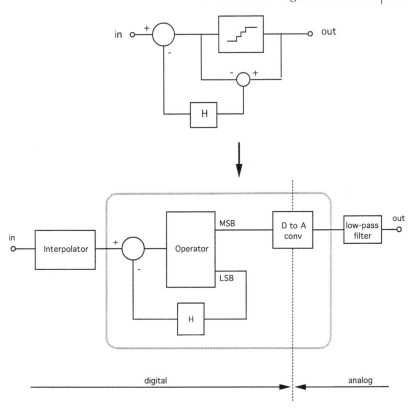

Fig. 7.35 A noise shaper suited for D to A Delta–Sigma converters (above), and practical implementation (below).

This is achieved thanks to a digital noise shaper whose purpose is to shift quantization noise as much as possible outside the baseband like in the A to D converter. At the noise shaper output, the word length of the oversampled data controlling the low resolution output D to A converter is shortened substantially. Since the *OSR* can be very large, a simple low-pass analog filter like a third order RLC network serves to filter the data from the reduced resolution D to A converter.

The actual noise-shaping algorithm is achieved by means of a non-linear feedback loop as in A to D Delta–Sigma converters. However, unlike these, digital data are processed instead of analog. The accuracy problems inherent to analog circuitry are thus avoided. Therefore the noise shaper may take the form of the circuit shown in the upper part of Fig. 7.35. In this circuit the quantization noise is sensed directly by subtracting the output from the input of the quantizer. The difference is fed back to the input through the digital loop filter that serves the same purpose as the one in A to D converters. Computing the difference is simplified because the words delivered by the noise shaper are split simply into two fields, an *MSB* and an *LSB* field. The *MSB* field word controls the reduced resolution D to A converter whereas the *LSB* field word is fed back through the loop filter 'H' as is illustrated in the lower part of Fig. 7.35.

In one of the first 16-bit D to A Delta–Sigma modulators (Goedhart *et al.* 1982), the *MSB* field counted 14 bits whereas the *LSB* field consisted of only the two *LSB* bits. Delta–Sigma was advocated only to compromise the lack of accuracy of the dynamic current matching D to A converter considered in section 3.2.2. Lower resolution D to A converters are currently advocated nowadays. Schouwenaars *et al.* (1992) takes advantage of a 6-bit D to A converter that exploits dynamically refreshed current mirrors like those described in Section 3.3.4. The converter consists of fourteen current sources divided into two equal groups that deliver positive and negative currents. Every input word picks up groups of 1,2 and 4 current sources, positive or negative, and puts them in parallel. Since no current source is activated for the zero level, offsets are zeroed automatically. Another approach to counteract the impairments of the D to A converter is to use unit-elements randomization. This is similar to the techniques that were used in order to lower the harmonic content of the D to A converters in the feedback loop of the noise shapers considered earlier.

Fast A to D converters   8

Many applications in the rapidly expanding fields of telecommunications, video, medical imagers, radar and network analyzers require converters whose sampling rates range from a few MS/s to hundreds of MS/s. Some applications, like digital oscilloscopes, require rates as high as 1 or 2 GS/s. Such large sampling rates require *parallelism* to perform the conversion in a single clock cycle. The D to A scaled converters considered in Chapters 2 and 3 are typical parallel devices. Their input code word controls the whole set of binary weights that construct the analog output, regardless of the number of bits. Therefore, they achieve sampling rates exceeding hundreds of MS/s. However, analog to digital conversion is more difficult. Neither the successive approximations of Chapter 4 nor the algorithmic converters of Chapter 5 are amenable for they require as many clock signals as bits. The only converters that truly operate in a single clock step are *flash* converters. They are the A to D counterpart of scaled D to A thermometer-coded converters. But unlike the latter, the power consumption increases very rapidly when the resolution surpasses 8 bits. Mixed architectures that combine small resolution flash converters and algorithmic architectures are able to offset the power budget at the expense of two or three clock cycles. These are *subranging*, *pipelined*, and *folding* converters and are reviewed in this chapter after detailed examination of the advantages and disadvantages of flash converters. Though slower than flash converters, they serve for many applications whose sampling rates are no larger than a few tens of MHz.

8.1. Flash converters

Figure 8.1 shows the block diagram of the generic A to D fully parallel *flash* converter. The input signal is simultaneously compared with a set of DC voltages implementing the analog counterpart of all the N-bit code words the flash converter may output. These are delivered by a multi-tap divider formed by 2^N equal series resistors. The comparators bank outputs a string of bits, which is the *thermometer-coded* input. Since this is a very redundant representation, it must be replaced by the equivalent *N*-bit binary-coded word. Code conversion is done in the digital encoder after the comparators bank. Here the thermometer-coded word is changed into an address that controls a

184 | Fast A to D converters

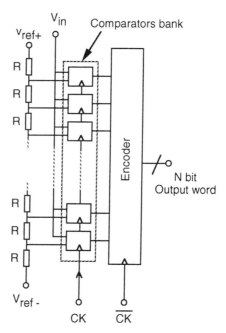

Fig. 8.1 The generic flash converter.

ROM memory, which outputs the desired binary-coded output word. The ensemble, comparators bank and encoder, forms a pipelined network that introduces two or three clock cycle latency, but does not affect the sampling rate.

Flash converters formulate a number of intricate challenges, the most demanding of which relate to the comparators bank. A large area and a lot of power are required, because the number of comparators increases exponentially with the resolution. The first ever reported flash converter was a bipolar 8-bit device with 5 MHz bandwidth requiring 2.5 W power consumption (Peterson 1979). This is still rather modest compared with the 12 W required by the 8-bit, 250 MHz bandwidth converter reported a few years later in Peetz (1986). Although both achieve the same resolution, the power consumption of the second is much larger for its comparators are much faster, and consequently require more currents. Much better figures have been achieved since then. Bult and Buchwald (1997) describes a 10-bit 50 MS/s flash converter that consumes only 70 mW. One of the highest sampling rates reported is the 2 GS/s 6 bit converter described by Wakimoto *et al.* (1988) which consumes only 2 W.

Flash converters are considered full-Nyquist converters when their bandwidth reaches half the sampling rate. Often converters exhibit bandwidths that are less than half the sampling rate. For instance, Akazawa et al. (1987) describes a bipolar 8-bit 2.7 W converter whose bandwidth reaches 100 MHz, while its sampling rate is 400 MS/s. Therefore, a clear distinction should be made between sampling rate and bandwidth to avoid confusion, especially when slew-rate limitations are considered (Mangelsdorf 1990). The present trend is to define not one but several bandwidth figures versus the magnitude of the input signal (e.g. small and large signal bandwidths), or to replace bandwidth considerations with other considerations, e.g. SFDR versus sampling frequency.

A distinctive feature of flash converters when compared with other A to D converters is the irrelevance of a sample-and-hold front-end. There is no need to freeze the input signal, as the thermometer code is already a binarized representation of the input. This is an interesting remark for comparators are always much faster than S.H.s since they ignore the Op Amps limitations. However, the discussion should not be restricted to the irrelevance of the S.H. front-end. Non-linear distortion is another problem that favors S.H. front-ends. Flash converters exhibit very large parasitic input capacitance, as a result of their large number of comparators. A figure as high as 300 pF was reported for the 8-bit converter described by Peterson (1979). Today this is usually closer to 20 (Akazawa et al. 1987) and 1 pF (Bult and Buchin 1997), though this is still a large capacitance given the frequency range of the input signal. The input capacitance exhibits strong non-linear contributions that are the result of the input comparators. Since the large frequency of the input signal produces large input currents, severe harmonic distortion may result from the non-linear input capacitance combined with the impedance of the input signal generator. When an S.H. front-end is used, the signal input to the flash converter is a DC signal and the distortion problem is shifted to the S.H. circuit, which is another source of harmonic distortion.

Clock jitter is a serious potential source of non-linear distortion as previously dealt with in Section 4.1. Small irregularities in the clock rate attributable to phase noise are the equivalent of small phase shifts of the input signal (Shinagawa et al. 1990). Noise in the comparators produces similar effects. Besides clock jitter, the layout also causes problems due to the differential delays that it is likely to introduce. Indeed, the clock and the input signal may travel over large distances to reach every comparator. It is not possible to maintain equal time to reach every comparator. Even though these times are inherently small, typically tens of ps, they affect the signals 'seen' by the comparators by position-dependent phase shifts. When the resolution increases, the problem gets even worse because the sampling window must be narrower and the distances increase at the same time. The only sensible strategy to limit the impact of position-dependent time delays is to configure all the paths to the comparators along a comb-type distribution.

In addition to the comparators, the reference scale is another source of potential problems. The divider generally consists of a long strip of resistive material, either aluminum or polysilicon. Aluminum exhibits a small resistance per square and therefore consumes a lot of power since the reference must sustain the full reference voltage. This drawback is not observed when polysilicon is used, because its resistance is normally a few tens of Ohms per square, but the reproducibility is less satisfactory. Laser trimming can be contemplated but its use is restricted due to mismatches from thermal stressing. An attempt to overcome the problems associated with the variable impedance of the reference scale and kickback noise for sampling rates as high as one GS/s is reported in Schmitz *et al.* (1988). It is based on performing the magnitude discrimination directly by means of variable threshold voltage comparators. This is only conceivable in an advanced MOS technology where independent threshold voltage adjustments can be done through local ion implantation. However, the achievable tolerances limit this technique to resolution not exceeding 4 bits.

8.1.1 Impact of comparator impairments

The core of any comparator is a clocked flip-flop. During the sampling or acquisition phase, the positive feedback loop of the flip-flop is annihilated so that its inverters assume the same state without switching. This is the so-called *metastable* state of flip-flops. While this is taking place, the difference between the input signal and the reference applied to the comparator is sensed by means of a low gain wideband differential amplifier whose output produces a small debalance among the two inverters. Once the actual sampling is launched, the positive feedback loop is restored and the flip-flop assumes the state induced by the debalance during the acquisition phase. When switching is completed, the state of the flip-flop represents the sign of the difference between the input signal and the reference applied to the comparator.

In the bipolar implementation shown in Fig. 8.2, the differential pair Q1 and Q2 tracks the input signal while the flip-flop is disabled. When the state of the clock changes, the flip-flop amplifies the difference sensed before the clock transition by the differential pair. The behavior of the MOS version shown below is similar. Transistor Q7 shorts the output of the flip-flop to counteract the positive feedback loop formed by Q8 Q9 and Q3 Q4. When the clock goes high, Q7 is blocked and the flip-flop switches. In the bipolar circuit, the sensing is done by switching the tail currents of the differential pair and flip-flop, while in the MOS version the differential pair is cut-off by transistor Q6.

The reference scale linearity and the comparator offsets are two factors defining ultimate discrimination. Mismatch errors in the resistive divider may be the consequence of plasma-induced defects produced during the fabrication

Flash converters | 187

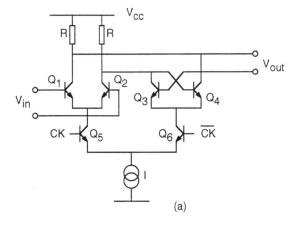

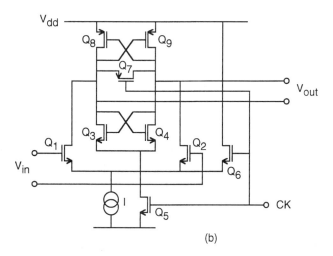

Fig. 8.2 Clocked comparator circuits: (a) bipolar, (b) CMOS.

process, such as thermal effects and radiation gradients. The correspondent errors have a direct impact upon the *INL* but cause negligible *DNL* distortion and no non-monotonicity errors. Offsets are a serious problem that becomes crucial when the resolution exceeds 8 bits in bipolar circuits and 6 to 7 bits in MOS. The flip-flops offsets are quite large indeed, typically 50 mV, but they are divided by the gain of the differential pair, which usually ranges from 5 to 20 mV. The challenge is to control carefully the offsets of the differential pairs by proper sizing. In the following, the bipolar and MOS technologies are considered separately.

In bipolar technology, offsets are controlled by the sizes of the emitters of the differential pairs. The impact of patterning errors, like underetching, get less critical as the sizes of the transistors increase. The trade-off size versus offset actually sets the limit of the achievable resolution. A flash converter with a 2 V reference aiming at a resolution of 10 bits, implies offsets no larger than 1 mV. Such a tight figure requires very large emitter sizes. When the resolution is lowered from 10 to 8 bits, the offset tolerance is multiplied by four while the number of comparators is divided by four. In practice, bipolar flash converters become unpractical above 8 bits. The input base current is another cause of offsets. It produces unwanted voltage drops across the resistive reference. If the impedances 'seen' from the comparators were all the same, a constant offset would be experienced; however, this is not the case for the resistance fluctuates from a minimum at the extremities to a maximum in the middle. Consequently, bias current from the differential pairs produces a kind of voltage bow along the reference divider that affects the *INL*. Lower resistances restrict their impact but the power consumption of the divider becomes more important.

Threshold variations are the main contributors to offsets in MOS converters (Pelgrom *et al.* 1994). They are larger than in bipolar transistors and reduce the resolution to 6 to 7 bits unless compensation techniques are used, but this may affect the sampling rate. An interesting approach to reduce the offsets between adjacent differential amplifiers has been proposed by Kattmann and Barrow (1991). The output of each comparator is no longer determined only by its own decision, but by that of its neighbors as well. Since offsets are supposedly randomly distributed, they are likely to compensate one another to some extent. The implementation is as follows. Every inverter output is linked to its two neighbors by means of coupling resistors. The resultant resistive chain embodies the mutual influence of 2 or 3 consecutive differential pairs. The smaller the resistors, the larger the interaction. This reduces the gain of the differential pairs however, so a compromise must be found that maximizes the desired offset averaging while maintaining the gain at a reasonable level. When the optimum is reached, threefold reduction of the *DNL* is obtained, which is equivalent to a resolution improvement of 1.5 bit. A gain of 4 and 2 bits, respectively, regarding the *DNL* and the *INL* is possible if the load resistances of the differential pairs are replaced by current sources. This is done in the 10-bit flash converter reported in Bult and Buchwald (1997). Each pair is terminated by a differential load consisting of cross-coupled current mirrors as shown in Fig. 8.3. Since the differential impedance is infinite, coupling through the horizontal resistors is maximized. In Bult and Buchwald (1997), five pairs influence each other but the gain of each pair is still about five. The random offsets are so much reduced that smaller, even minimum, size transistors become acceptable. An additional interesting feature of the circuit of Fig. 8.3 is its low common mode impedance waiving the need for common

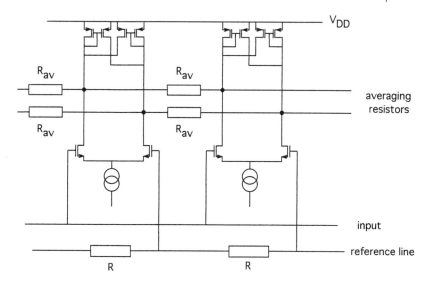

Fig. 8.3 In the averaging technique proposed by Bult and Buchwald (1997), current mirror loads maximize the interdependence between adjacent channels in order to average out the random offsets of the differential pairs.

mode regulation. This flash converter has been implemented in a 0.5 μm CMOS technology. Its area is only 1 mm² and the input capacitance 1 pF. The power consumption is 170 mW for a sampling frequency of 50 MS/s.

Besides static impairments, a number of dynamic effects must also be considered. Every differential amplifier in front of a comparator 'sees' an input signal whose magnitude may vary between large limits. Thus, the amplifier operates in the non-linear as well as the linear regions. It has been shown that in order to keep the third harmonic distortion of 250 MHz sine waves below −60 dB, the bandwidth of the differential stage should reach 5 GHz! The reason is that the bandwidth/non-linearity combination introduces a variable time lag, known as the *signal dependent delay* described in van Valburg and van de Plassche (1993). Another dynamic effect that must not be overlooked is the impact of transient currents flowing through the input terminals of the comparators. These produce dynamic offset known as *kickback* noise, which is associated with the parasitic capacitances coupling the input terminals to the clock signal. They cause short, very sharp voltage perturbations in the divider sensed by every differential pair during switching. The problem is complicated by the fact the impedance of flip-flops undergoes a sharp minimum during switching. These transient currents offset the comparison levels at the worst possible moment. Non-switched differential pre-amplifiers separating the input from the flip-flop are a sensible way to reduce kickback noise.

8.1.2 Metastability, encoding errors and digital correction

The time allocated to comparators to decide whether an input is larger or smaller than the reference is determined not only by the bandwidth of its differential input amplifier but also by the time constant of the resolving flip-flop. In Fig. 8.2, this time constant is determined by the transconductances of Q3 and Q4 (plus Q8 and Q9 in the lower figure) and the parasitic capacitances at the collector or drain terminals. Although this time constant is very small, typically a few hundreds of ps or less, the switching time of the flip-flop may vary considerably. The flip-flop outputs a signal that is a replica of the input debalance increasing exponentially. When the debalance is large, switching is fast, but when contrary, the time to reach a clear distinct state may be quite long. Switching times increase in a similar way to the log of the reciprocal of the input debalance. Thus, when the input debalance is very small, the flip-flop may remain indecisive for a time exceeding the allocated sampling time window and lead to a wrong decision. This phenomenon, called *metastability*, is unavoidable, hence some action must be taken to reduce its consequences.

One of the consequences may be the introduction of a 'bubble' in the thermometer-coded output of the comparators, which may seriously endanger the final binary-coded output. Most encoders consist of two layers of hardware. The first transforms the thermometer code into an intermediate 1-out-of-N code word with zeros everywhere except at the transition defined by the input. The second layer, which may be assimilated to a read-only-memory, is controlled by the output of the first. Figure 8.4(a) shows a simplified illustration of the first layer as well as input and output signals. The output data from the comparator are shown on the left and the code word feeding the second layer on the right. It is clear that the bubble produces an undesired '1', two levels below the transition. Thus, two words are being addressed at the same time in the second layer. This can lead to an output code that differs considerably from what it should have been. The result may be very damaging to the spectral signature.

The circuit shown in Fig. 8.4(b) illustrates a strategy that can be used to counteract the consequences of single bubbles. It uses three instead of two consecutive comparator outputs to determine whether a zero is a bubble or not. The circuit does not correct the more unlikely situation where two consecutive bubbles are experienced simultaneously. More sophisticated decoders are capable of correcting this error. The solution is supplemented mostly by code transformations that minimize the difference between adjacent coded words. In the 80 MS/s converter reported by Portmann and Meng (1996), a pipelined decoder is described that reduces error rates from 10^{-4} to less than 10^{-12} errors/cycle.

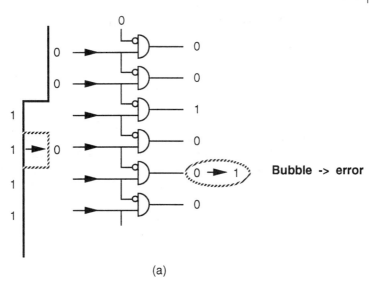

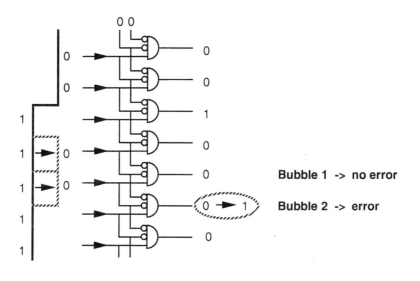

Fig. 8.4 Bubbles caused by metastability (a) can be counteracted by means of additional logic (b). Adjacent bubbles are not corrected however in (b).

8.2. Subranging, two-step, multi-step and pipelined A to D converters

The large number of comparators needed by flash converters calls for alternative solutions. A sensible compromise is offered by *subranging, two-step, multistep* and *pipelined* converters, which drastically reduce the numbers of comparators by performing the A to D conversion in two or more steps. In the two-step converter, the unknown is coarsely quantized by a reduced resolution flash converter. This coarse code determines the segment that encompasses the input signal. Fine conversion take place within the selected segment immediately after. This spectacularly reduces the number of comparators. A 10-bit converter consisting of two five-bit wide sub-converters, requires 62 comparators instead of the 1023 for full flash conversion. This represents not only substantial savings as far as area and power but also a sensible way to reduce the input capacitance. However, at least two clock cycles are required. Thus the sampling rates are hardly competitive with those of flash converters. Tens of MS/s can be achieved and the resolution may exceed 10 bits.

8.2.1 Subranging architectures

Two-step converters can be implemented by splitting the reference scale into a coarse- and a fine-section. During the coarse conversion, the input signal is digitized against the coarse scale. The output code is used to select the set of fine references that encompass the input signal. Once the fine code is resolved, coarse and fine codes are concatenated to obtain the actual output code word. In the example shown in Fig. 8.5, the reference scale is implemented by means of a long strip of resistors (Dingwall 1985). The coarse scale selects an appropriate sub-set of resistances that becomes the reference scale of the fine flash converter below. If M bits are resolved during the coarse conversion and P during the fine, the total number of comparators equals $(2^M + 2^P - 2)$. Note that the offset tolerances of all comparators must be the same as those of full flash converters. The name *subranging* is given to this kind of converter, which may be viewed as the multi-bit extension of successive approximation converters.

The number of comparators can be further reduced if the same set is used twice to perform the coarse and fine conversions successively. This is done in the 8-bit subranging converter described by Hosotani *et al.* (1990), which requires only 15 comparators. The converter uses the same 4-bit flash converter twice. To circumvent the increased conversion time due to switching and zeroing the bank of 15 comparators, the actual converter consists of two interleaved identical subranging converters. The sampling rate reaches 20 MS/s whereas the power consumption is only 50 mW and the input capacitance 10 pF.

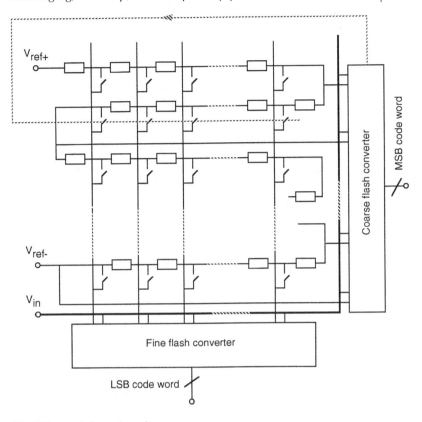

Fig. 8.5 A resistive subranging converter.

8.2.2 Two-step converters

Instead of changing the reference scale like in successive approximation converters, one may also modify the input signal during the course of the conversion as in algorithmic converters. This occurs in the *two-step* converter shown in Fig. 8.6. Here, the difference between the input and the analog counterpart of the coarse code word is computed first. After subtracting the analog counterpart of the coarse code from the input, the difference, or *residue*, is digitized by the fine flash converter that delivers the fine N2-bit code. The output word is the concatenation of the N1-bit coarse code delivered by the first flash and the N2-bit fine code.

The algorithm illustrated in Fig. 8.7 considers a two-bit coarse A to D converter. The vertical bar in the third segment represents the input signal. After the analog counterpart of the coarse code is subtracted from the input, the small vertical vector below is obtained. Residues are represented by the

194 | Fast A to D converters

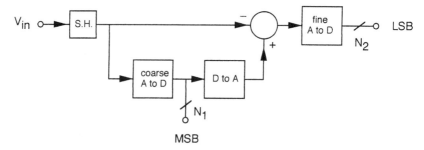

Fig. 8.6 In two-step converters, the analog counterpart of the *MSB* code is subtracted from the input and the residue is applied to the fine converter.

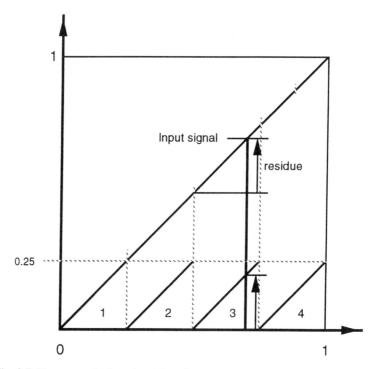

Fig. 8.7 The sawtooth shaped residue of a two-step converter.

sawtooth shaped waveform below, which represents also the quantization noise of the two-bit first conversion.

As in subranging architectures, the main asset of the two-step architecture is the substantial reduction of the number of comparators that results from dividing the flash conversion into two steps. The main drawback lies in the extra time needed to calculate the residue and the concurrent need to implement a S.H. The fine conversion cannot start until the coarse analog-to-digital-to-analog conversion is completed. Since the associated delay generally exceeds the width of the sampling window, the input must be sampled and held. This introduces an Op Amp with its limitations associated to the existence of a dominant pole.

8.2.3 Recycling converters

An interesting architecture derived from the previous two-step converters is the *recycling* converter. The converter uses the same hardware, but instead of a second flash to resolve the residue, it recycles the amplified remainder to the input. A new string of M bits is output every cycle. After being concatenated, these represent the final output code word. Thus the fine converter shown in Fig. 8.6 is eliminated at the expense of an auxiliary gain stage that is needed to inflate the residue to match the dynamic range of the unique flash converter. The gain is fixed by the resolution of the A to D to A block. If M bits are resolved per cycle, the gain of the auxiliary amplifier must equal 2^M.

Recycling enables reducing the number of comparators without affecting the overall resolution by taking larger numbers of cycles, but the conversion time increases. The smallest number of comparators corresponds to a pair of single-bit A to D and D to A converters. The recycling converter is then a true algorithmic converter like those considered in Chapter 5. The A to D converter resumes to a comparator and the D to A to a switch that performs the conditional subtraction of the reference voltage $V_{ref}/2$, which represents the single-bit fractional approximation of the input. The interstage gain moreover equals 2. The similarity with algorithmic converters extends to the Robertson plot, as we will see later.

8.2.4 Pipelined converters

When more than two cycles are considered, *pipelined* converters offer a sensible compromise to counteract the lower sampling rates of recycling converters. A pipelined converter consisting of three identical M-bit blocks is shown in Fig. 8.8. Each block resembles the previous recycling converter, but its output feeds another block instead of being recycled. The peculiarity of this kind of architecture lies in the fact that each block operates on different data. While the residue from the first block is being transferred to the second, a new sample enters the first block and so on. Consequently, the conversion rate is determined by the time needed to transfer data from one block to the next, not from the input to the output of the converter. Therefore, pipelined converters allow to increase the resolution without increasing the conversion time nor by

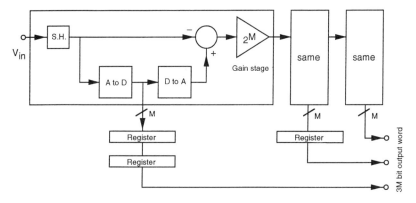

Fig. 8.8 In pipelined converters, cascaded stages resolve residues after they have been multiplied by the interstage gain. The partial code words are re-ordered by means of shift registers before concatenation.

sacrificing area and power consumption. A new degree of freedom is introduced for they capitalize both resolution per stage and number of stages. Note that one S.H. network per stage is required to memorize the data processed by each stage and that shift registers are needed to reconfigure the correct output data memberships.

The minimum number clock cycles pipelined converters require is two. The sampling rates currently range from a few MS/s to a hundred MS/s while the resolution ranges from 10 to 13, even 16 bits. Several examples are reviewed in the following.

8.2.5 Impact of imperfections

Current impairments that limit multi-step converter performance are the non-ideal transfer characteristics of the A to D and the D to A converters as well as the interstage gain error (Hadidi *et al.* 1992). To discern the impact of all possible sources of impairments, a generic converter is considered hereafter consisting of a non-ideal 2 bit coarse-stage followed by an ideal infinite resolution fine-stage. If the coarse-stage were perfect, all the residues would portray the ideal sawtooth waveform of Fig. 8.7. However, the case studies displayed in Fig. 8.9 show a number of possible impairments. These are *transition position* errors (left), *transition magnitude* errors (middle) and *gain* errors (right). Each plot displays a transfer curve shown above and the correspondent residue below. Only one type of error is considered per vertical frame. Transition position errors (left) are caused by uneven discrimination levels inside the flash A to D converter (*DNL* errors). The residue below shows the concurrent transition position displacements (plain lines) with

Subranging, two-step, multi-step and pipelined A to D converters | 197

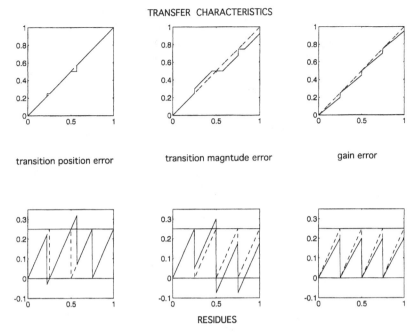

Fig. 8.9 Multi-step converters are subject to a number of impairments like transition position errors (left), transition magnitude errors (middle) and interstage gain errors (right). Their impact on the transfer characteristic and residue is shown respectively above and below.

respect to the ideal (dashed lines). Neither transition magnitude errors nor gain error are considered, because the D to A converter and the interstage gain are assumed to be ideal. Several distinct effects are exemplified. In the first, the transition occurs too early, in the second, too late, and in the last, just in time. The consequences are illustrated by residue overflows and underflows, respectively. After being multiplied by four, the residue overrides the dynamic range of the flash converter. The consequences are illustrated as the flat segments in the transfer curve, which cause large positive and negative *DNL* errors.

Transition magnitude errors (middle) are caused by the D to A converter impairments. Neither the transition positions nor the slopes are affected, but the magnitudes of the transitions do not entail the ideal 0.25 step height. Once again this causes over- and under-flows. The *INL* and *DNL* are seriously damaged, for entire segments of the transfer characteristic are now offset.

Transition gain errors lead to the situation shown on the right. Note that the errors from the D to A converter and the intermediate gain stage only affect the fine conversion, never the coarse conversion. Thus their impact depends on the number of bits M resolved per cycle. When M is large, say 4 or 5, the tolerances regarding the transition magnitude and interstage gain are reduced. A sigma of 1% affecting the D to A converter unit-elements as well as interstage gain does not seriously endanger the accuracy of a 10-bit converter. Conversely, when M is 2 or 1, the tolerances concerning the same elements must be considerably narrowed.

8.2.6 Correction strategies

This section reviews techniques that counteract the imperfections described previously. A two-step architecture is considered but the resolutions of the first and second conversions are assumed to be equal to ease the description of the correction mechanisms. First, let us consider the impairments of the flash converter. As mentioned earlier, these are the result of comparator offsets as well as mismatches affecting the reference divider. The two plots of Fig. 8.10 display the data output after the first and second flash conversions, considering 2-bit quantization steps. The plot above relates to the ideal converter and the one below, to the same device after introducing an error in the first flash. Transitions should be located at 0.25, 0.50 and 0.75 if we consider a dynamic range from zero to one. In the plot below, the mid-transition of the first flash has been shifted from 0.50 to 0.65. Consider an input equal to 0.60. After coarse conversion, code 01/00 is issued instead of 10/00 (the slash symbolizes the separation between coarse and fine codes after concatenation). Since the D to A converter is supposedly ideal, the analog counterpart of the erroneous code 01 is subtracted from the input. After multiplying the residue (0.6 − 0.25) by four, the second flash overflows since 1.40 lies out of its dynamic range. Suppose the dynamic range were twice as large so that the extended output code 0101 could be added to the wrong code 01/00 issued by the first flash conversion. The result will now be correct. Of course the second conversion, which is not ideal, introduces new errors that further propagate, but since their impact is divided by the interstage gain, the more bits resolved per cycle, the smaller the error will be.

Figure 8.11 shows the flash converter output data delivered by a recycling converter that uses a two-bit flash converter with severe transition position errors. The data after the first and second cycle are shown as dashed line curves whereas the converter transfer characteristic is shown as plain lines. Outside the dynamic range, the flash converter is bound to plus and minus one. The impact of transition position errors is easily spotted as is illustrated and the saturation of the fine data which produce a number of missing codes or strong *DNL* errors. In the lower plot, the dynamic range of the flash converter has been extended. As a result, the impact of the transition position errors is substantially attenuated. There are no missing codes. Only the *DNL* is affected but still little.

Subranging, two-step, multi-step and pipelined A to D converters

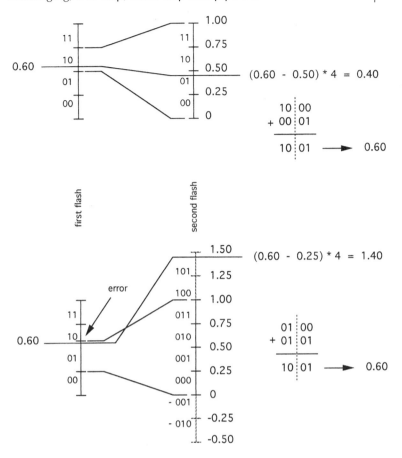

Fig. 8.10 Transition position errors are corrected digitally if the dynamic range of the flash and D to A converters are extended.

Since the dynamic range of residues is bound in practice by the power supply, a slightly modified scheme is generally adopted. The resolution of the flash converter is augmented by one bit and its dynamic range is kept unchanged.

Magnitude transition errors and intermediate gain errors are more arduous to correct. A number of specialized papers deal with this problem. In general the correction strategy is based on calibration procedures in the digital domain that take advantage of signal processing algorithms. A correction scheme called interstage gain error proration is described in Lee and Song (1992). Another, known as background digital correction is described in Shu et al. (1995) and Kwak et al. (1997).

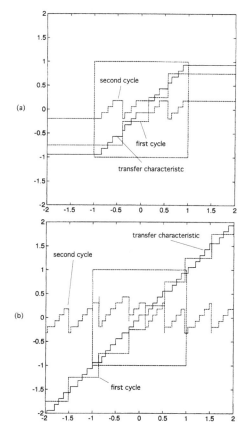

Fig. 8.11 Illustration of the residues (dashed lines) and transfer characteristic (plain line) of a 2-bit, two-steps converter having strong transition position errors. Before digital correction above, after below.

8.2.7 Examples of multi-step and pipelined converters

Combining a flash converter with a switched capacitor D to A converter leads to a recycling converter architecture that is very appealing. Figure 8.12 shows the 10-bit 15- MS/s recycling converter described by Song *et al.* (1990). In this device, the residue calculation is entrusted to a modified version of scaled D to A converters, known as MDAC, or Multiplying D to A converter. The flash converter recycles the residue to the MDAC. Transition position errors inherent to the flash converter are corrected digitally and errors related to the MDAC mismatches are negligible since their impact is divided by a large interstage gain of 32.

Subranging, two-step, multi-step and pipelined A to D converters

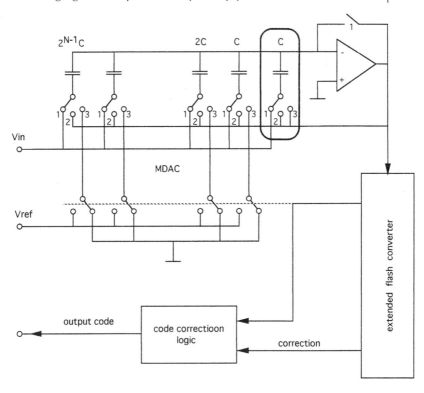

Fig. 8.12 The recycling two-step converter described by (Song et al. 1990).

The actual residue calculation is described hereafter. During the sample-phase (noted by 1), the input signal is applied to all the capacitors connected in parallel while the Op Amp is reset. During the hold phase (2), the full set of capacitors closes the feedback loop around the Op Amp to complete the S.H. sequence. The output voltage of the Op Amp is then a replica of the input voltage. Since it is applied to the flash converter, the latter outputs the coarse code of the input signal. Residue calculation can now take place (3). The bottom plates of all the capacitors except the last one, which is the only one remaining across the Op Amp, are connected either to the ground or to the reference V_{ref} depending of the outcome of the flash conversion. The charge stored in the capacitor's bank now represents the analog counterpart of the output code of the flash converter. Since it previously stood for the input signal, the departure is transferred to the integrating unit-capacitor around the Op Amp. Hence, the MDAC outputs the residue multiplied by the interstage gain, and the output of the flash converter is now the fine code.

This converter was implemented in a one μm CMOS technology (Song *et al.* 1990) and displays a power consumption of 250 mW and an input capacitance of 3.2 pF. Since the Op Amp is supposed to fulfill two distinct functions, its design is subject to very demanding specifications. During the sample mode it implements a non-inverting unity gain amplifier and during the residue mode, it becomes an inverting amplifier with a gain of nearly 30 dB. The converter resolution and sampling frequency set the specifications of the Op Amp. Hence, the gain and transition frequency of the latter are very demanding. The accuracy reached during the sample and residue modes requires loop gains of 45 and 72 dB, respectively. Settling times determine the transition frequencies, but this gives rise to an intricate problem. A settling time of 5 ns in the unity-gain configuration implies a transition frequency of nearly 200 MHz, which is feasible. A settling time of 20 ns in the residue mode yields a transition frequency of 32 times 50 MHz or 1.6 GHz because of the large interstage gain when a first order Op Amp is contemplated. This is too large a bandwidth and actually unnecessary since eventual poles below 1.6 GHz do not impair stability, as long as they don't get below 50 MHz. One can take advantage of this by splitting the Op Amp into two cascaded sections. In the unity gain configuration, a low gain (45 dB) wideband (200 MHz) single stage cascode pre-amplifier is used. This amplifier is preceded by an auxiliary amplifier to boost the overall gain until 72 dB in the residue mode that is bypassed in the unity gain mode.

Recycling converters are easily changed into pipelined converters. Instead of being recycled, the residue is merely fed to the next stage. The architecture is very straightforward for every MDAC already has its own built-in S.H. One of the first multi-bit pipelined converters of this kind was the 5 MS/s, 180 mW, 9 bit device described by Lewis and Gray (1987) that consisted of four, 3-bit wide, stages. Much effort has been devoted since then to design high performance pipelined converters. Present pipelined converters display sampling rates in the range of 20 to 100 MS/s while their resolutions are mostly between 8 and 10 bits. The power consumption becomes quite large when the sampling rate exceeds 50 MS/s. Figures between 800 mW and 1.1 W have been reported. A notable exception is the 250 mW figure quoted in Krishnaswamy *et al.* (1997), which uses low power design techniques and reduces the number of Op Amps from 7 to 3 by sharing the same amplifiers between adjacent stages. Fast converters (Kazuya *et al.* 1993) achieve 100 MS/s sampling rate with two stages. Kim *et al.* (1997) achieves the same sampling rate using a 4-bit front-end that distributes the residue to several identical time interleaved fine channels. In the 75 MS/s converter described by Colleran and Abidi (1993), the 4-bit first stage is followed by a 7-bit folding converter (see Section 8.3). A 50 MS/s, 10-bit, 900-mW converter is described by Yotsuyanagi *et al.* (1993). Pipelined converters in the 20 MS/s range currently reach 10 bits. Their power consumption is substantially

Subranging, two-step, multi-step and pipelined A to D converters | **203**

smaller, such as 35 mW quoted by Byunghak and Gray (1995) and 30 mW by Kusumoto *et al.* (1993). Low sampling rate pipelined converters aim for higher resolutions, from 10 to 16 bits. Their sampling rates are generally between 500 kHz and 5 MHz. In most of these, large numbers of stages are implemented while the resolved bits per stage are small. Small resolutions per stage reduce the bandwidth specifications of the MDAC Op Amp, but are more demanding regarding accuracy. Generally the converters require a smaller area despite a larger number of stages. Thermometer-coded MDACs advantageously replace binary-coded MDACs, *all the more* the flash converter delivers thermometer-coded data naturally. Stage resolution optimization is a problem dealt with in Lewis (1992). When a single bit is resolved per stage, the pipelined converter becomes an algorithmic converter. Matching problems regarding the flash and MDAC are ignored as they involve a comparator and a switch, respectively. However, offsets and intermediate gain errors become much more critical. In the pipelined 15-bit, 1 MS/s, described in Karanicolas and Lee (1993), the interstage gains of all 11 first stages are corrected digitally. Each gain is set slightly below two before calibration starts according to the correction algorithm described in Chapter 5. The 16-bit, 550-mW, 1.25 MHz signal bandwidth converter described in Brooks *et al.* (1997) combines Delta–Sigma and pipelined techniques for the front-end and the ensuing single-bit stages, respectively. Another 16-bit, 200 mW, converter is described in Mayes and Chin (1996). Its accuracy is checked by means of an embedded microcontroller that performs the calibration. As far as low-power power consumption, Yu and Lee (1996) describe a 12-bit, 5 MS/s pipelined converter requiring no more than 30 mW.

8.2.8 RSD multi-step converters

The RSD algorithm lends itself also to the implementation of multi-step recycling and pipelined converters that generalize the features discussed in Chapter 5, namely:

1. The *DNL* and *INL* characteristics ignore the Op Amp offset.

2. The transition position errors of the flash converter are compensated while the conversion goes on.

3. One additional bit is gained.

As discussed in Chapter 5, RSD words make use of numerals that take three values, $+1$, 0 and -1, instead of the usual 0 and 1. The M-numeral code words are redundant by essence; identical data can thus be represented by different words. This provides room for correction of the errors affecting the flash converter.

The main difference between RSD and binary-coded converters lies in the reference scale and coding of the flash converter. The reference divider comprises $(2^{(M+1)} - 1)$ resistors that are all equal except the two end resistors, whose magnitude is 1.5 times larger. For example, suppose M is one and consider a dynamic range between −1 and +1. The divider consists then of three resistors, 1.5 R, R and 1.5 R. These define two comparison levels, respectively −0.25 and 0.25, and the output takes three values, +0.5, 0 and −0.5, all representable by a single numeral. This is the RSD algorithmic converter of Chapter 5. Now suppose M equals 2. The divider now comprises seven resistors, which define six comparison levels, respectively −0.625, −0.375, −0.125, 0.125, 0.375 and 0.625, and seven quantization levels described by two-numeral words.

The Robertson plot considered in Chapter 5 may be extended to RSD recycling converters as well. The two examples shown in Fig. 8.13 relate to a converter for which M equals 3. The plot above concerns the ideal converter; the one below the same device after a number of transition errors are introduced. Trajectories are constructed in the same way as in Chapter 5. The trial number considered in the figure is 0.91234. After bouncing against the right-hand most gain line, the trajectory enters in the region that is common to the two orthogonal residue characteristics. Once inside, the trajectory stays there forever.

RSD converters are insensitive to transition position errors as cited in Chapter 5. It has been shown that changes in the two comparison levels do not modify the reconstructed output data, nor compromise the *INL*. Only the width of steps varies so that the *DNL* is affected slightly, but within the one-half *LSB* tolerance. As long as comparison levels don't overlap, the width of each step remains bound within plus or minus one-half *LSB*, regardless of the resolution. The same property applies to higher order RSD converters. Codes change but still represent correct data due to the redundancy. In the lower part of Fig. 8.13, a lot of transition position errors were introduced through resistance mismatches and comparator offsets whose standard deviations were equal to 0.05 and 0.02, respectively. Although the trajectory is drastically modified, the reconstructed RSD data are identical. The auto-correction holds true in fact as long as the trajectory does not hit a gain line outside the square comprised between plus and minus one.

As an illustration of the above, Fig. 8.14 displays the *INL* and *DNL* characteristics of the converter considered in the lower plot of Fig. 8.13 after three cycles to yield a resolution of (3.3 +1) or 10 bits. The two plots clearly illustrate the impact of the transition position errors, which control the width of the steps. However, none does jeopardize the *INL* and *DNL* characteristics. The transition position errors produce erratic length modulations of the steps. As in algorithmic converters moreover, the levels may change during conversion without impairing the overall accuracy. The noise generated by the flash converter comparators does not affect the output data consequently. Similarly, wrong decisions induced by metastability do not impair the accuracy either. Codes change but the data remain unharmed.

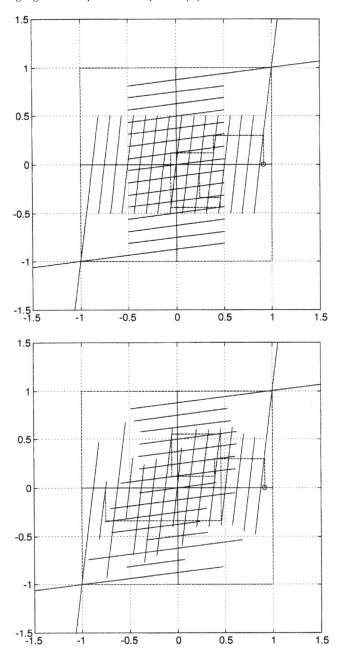

Fig. 8.13 Robertson plots of an ideal multibit RSD converter (above). The same plot after introducing transition position errors. The latter do not affect the reconstructed data.

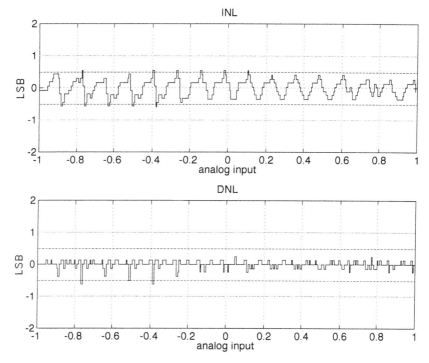

Fig. 8.14 *INL* and *DNL* characteristics of the ill-conditioned converter considered below in Fig. 8.13.

The auto-correction mechanism is similar to the correction algorithm described in Section 8.2.6, which inflated the dynamic range of the binary-coded converter to linearize the output. In the RSD converter, the redundant numerals take care of this.

The same rules as above apply to MDAC impairments, whether binary or RSD flash converters are considered. The signal-to-noise versus magnitude plot of Fig. 8.15 illustrates the same interesting feature that was mentioned in Chapter 5. The two *SNR* curves relate to data output by the converter considered above. The linear curve relates to a converter having transition position errors, but an ideal MDAC. The other curve that bends near the maximum, relates to the same ill-conditioned converter after introducing errors with a standard deviation equal to 0.005 in the MDAC unit-elements (and thus, in the interstage gain also). The only part of the curve that is affected by the MDAC impairments concerns large signals. Below −20 dB, there is no difference and the *ENOB* is the same regardless of the converters.

The spectral distributions shown in Fig. 8.16 relate to the same converters as those shown in Fig. 8.15. In the upper plot where the MDAC is ideal, the

Subranging, two-step, multi-step and pipelined A to D converters | 207

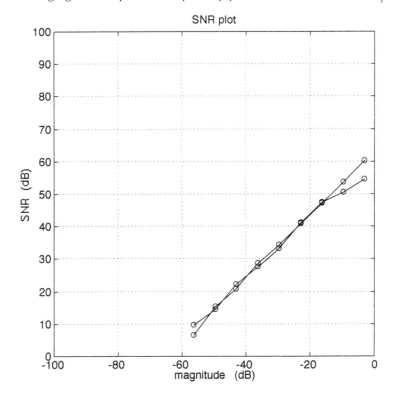

Fig. 8.15 The *SNR* of RSD converters (upper curve) is not affected by transition position errors, while magnitude transition errors bow the characteristic near saturation (lower curve).

spectrum does not uncover any impairment despite the transition position errors. In the lower plot extra low frequency noise and intermodulation products are clearly visible as a result of the MDAC impairments. MDACs using thermometer-coded weights instead of binary-coded weights may improve the harmonic content, provided the unit-elements are randomized as in multi-bit Delta–Sigma converters (section 7.10.2).

A 4 MS/s, 10-bit, 20-mW pipelined converter implemented in a 3 μm technology that uses the RSD algorithm is described in Ginetti (1992). To maintain the power consumption below 20 mW, the number of Op Amps was halved, swapping the same amplifier between consecutive stages. The same algorithm is also used in the 10-bit, 9 stage converter described by Lewis (1992), which was implemented in a 0.9 μm technology, achieving a 20 MS/s sampling rate for 240 mW power consumption. Byunghak and Gray (1995) achieve the same performance with only 35 mW. A 4.5 MS/s, 10-bit 20 mW

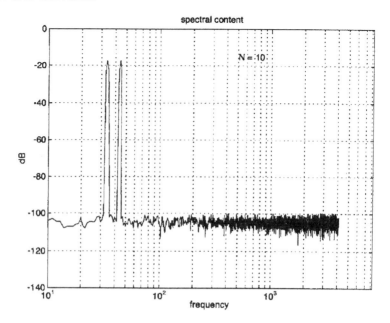

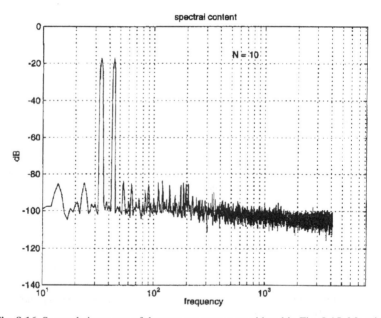

Fig. 8.16 Spectral signatures of the two converters considered in Fig. 8.15. Magnitude transition errors increase the low frequency noise floor and cause intermodulation (below) while transition position errors don't (above).

current mode version is described in Macq and Jespers (1994). All these converters use single, rather than multiple, numeral stages. The latter have not yet been reported, although a similarity with the multi-bit implementation described in Cline and Gray (1996) cannot be excluded.

8.2.9 Optimization of the S.H. switching sequence

Sample-and-hold circuits intended for fast A to D converters deserve mention for they must fulfill many conflicting requirements. Besides high speed, small area and low power consumption, they should introduce minimal harmonic distortion (Lin *et al.* 1991).

Techniques that minimize the signal dependent charge from the switches are particularly important. Generally, one tries to exchange signal dependent for constant charge injection; the latter being turned into a constant offset, which is ignored by the differential architecture of the converter.

In the circuit of Fig 8.17, the voltages across Cs+ and Cs− track the differential input signal V_{in} as long as transistors Q1 and Q2 are conducting. While this is taking place, the integrating capacitors Ci are reset, the Op Amp is shorted and its input terminals connected to ground by two small transistors Q7 ($W/L = 10$). Shortly before sampling, the latter are cut off. Sampling takes place when Q1 is cut off so that the capacitors Cs are open ended (the parasitic capacitances C_p are assumed to be very small). The injected charge from Q1 is split into two equal parts since both junctions of this transistor were zeroed before cut-off. These charges produce voltage steps across the input terminals of the Op Amp with equal magnitudes. They are ignored given the differential architecture. Now transistors Q2 are cut-off. Their charge, which is signal dependent, has no other way to go than to return to the input generator, with the exception of the small fraction absorbed by the parasitic capacitances C_p. Finally, the charge stored in the sampling capacitors Cs is transferred to the integrating capacitors Ci after switching 'on' transistor Q6.

Charge injection counter-measures like these are crucial when high frequency signals are being sampled. The currents flowing through the sampling capacitors Cs can be very large indeed. To minimize the voltage drops across the terminals of the series transistors Q1 and Q2, the latter must be very large ($W/L = 100$). Their contributions to charge injection are thus substantial, justifying the above compensation techniques.

8.3 Folding converters

Folding converters belong to a class of converters that operate like two-step converters; but contrary to these they don't require Op Amps. Hence, they surpass subranging and pipelined converters as far as speed. Sampling rates

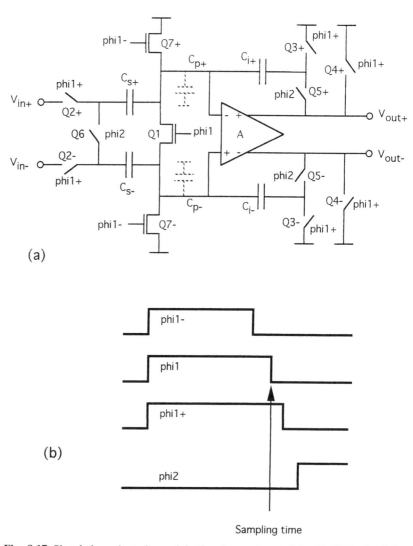

Fig. 8.17 Signal dependent charge injection is counteracted in this S.H. circuit by proper phasing of the clock signal and the use of a differential architecture.

reach currently a few hundred MS/s making S.H. front-ends purposeless. The resolution generally lies around 6 to 8 bits with the remarkable exception of 12 bits reported in Vorenkamp and Roovers (1997). Folding converters take advantage of high speed analog signal processors to avoid the bandwidth impediments related to Op Amps. Their name comes from the circuit computing the residue, which is described in the following.

8.3.1 The folding principle

Two generations of folding converters can be traced. The first is illustrated by the current-mode folder shown in Fig. 8.18 (Huijsing et al. 1993; van de Plassche and Baltus 1988). The circuit consists of several current sources connected to a pair of common load resistors through a set of common base transistors. When the input current I_{in} is zero, all diodes are unbiased, thus equivalent to open circuits. Suppose the magnitude of the input current increases monotonically. Current is first taken from the left current source. Since the emitter voltage of Q1 cannot change, the diode D1 remains blocked. Thus, the collector current of Q1 decreases while the input current increases, until I_{in} equals I_o and transistor Q1 is blocked. Now the emitter voltage increases to allow extra current to be drawn from the next current source through diode D1. The same mechanism repeats itself as far as transistor Q2, and so on. Because odd and even collector currents are injected in opposite resistors, the voltages across the load resistors exhibit positive ramps that alternate with horizontal segments. The differential output voltage V_{out} looks like the triangular waveform shown in Fig. 8.19. Although it differs from the curve shown in Fig. 8.7, the modified residue retrieves identical data. These enable assessing the fine code while the coarse code may be derived from the number of conducting diodes.

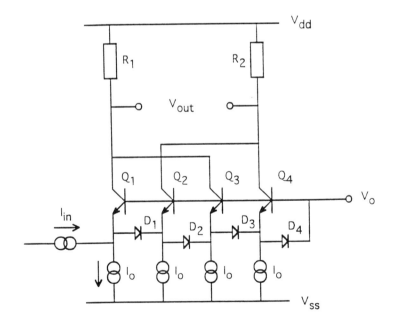

Fig. 8.18 An early current-mode folding circuit.

212 | Fast A to D converters

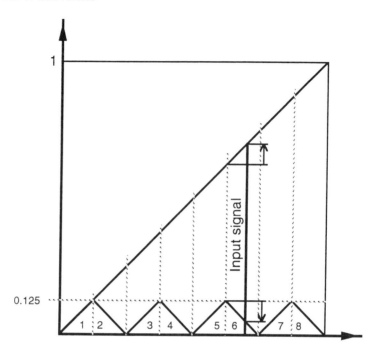

Fig. 8.19 Folding converters produce a residue that differs slightly from the one shown in Fig. 8.7.

The main difficulty concerning folding circuits lies in the very large bandwidth required to compute the residue. Triangular signals like those delivered by the folder encompass far reaching harmonics. The frequency of folded signals moreover is a multiple of the input signal frequency. Since the correspondent bandwidths are exceedingly large, the peaks of the triangular residue are currently rounded off. This causes severe errors and a lot of harmonic distortion. The only more or less distortion-free sections of the residue are the linear portions amid the peaks. It is this observation that is put to use in the folding converter described in Huijsing *et al.* (1993), which combines two converters with folders shifted by 90° as suggested in Fig. 8.20. The idea is to periodically switch from one folder to the other to avoid the peaks and exploit only the linear sections in between.

8.3.2 Folding and interpolation

Because the main drawback of the above converters lies in the bandwidth of the folder, the magnitude of the residues is given up for resolution in time. A second generation of folding circuits was introduced that takes

Folding converters | 213

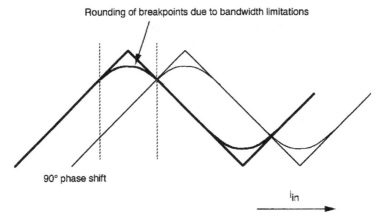

Fig. 8.20 Two folders shifted by 90 provide means to reduce the distortion caused by the finite bandwidth of the folder.

advantage of the zero crossings of the residue, instead of the magnitude. Since the shape of the residues is no longer important, folders with a smaller bandwidth become acceptable. The circuit consists of a bank of identical differential pairs that amplify the difference between the input signal and the reference scale as shown in Fig. 8.21. The drain currents are fed to a

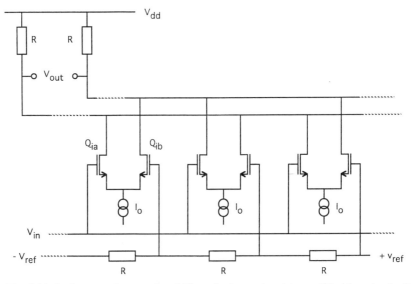

Fig. 8.21 In the second generation folders, the input signal is amplified by a bank of differential amplifiers whose outputs alternate.

common differential load after exchanging even and odd output terminals. When a ramp voltage is applied to this circuit, each pair delivers two currents, one evolving from zero to the tail current and another vice-versa. The summed output currents produce a differential voltage across the load that increases and decreases periodically. The result looks more or less like a distorted sinusoidal signal instead of the sawtooth displayed in Fig. 8.19. Its shape is controlled moreover by the contributions from neighboring pairs as illustrated in Fig. 8.22. Consequently the output looks like a rounded square wave when little or no overlapping takes place and a quasi-sinusoidal waveform when overlapping is taking place. Regardless, the aim is to produce regularly spaced folds.

Figure 8.23 shows the output of four folders that are shifted to divide the horizontal axis in four equal parts. Consider the input signal highlighted by the vertical line that intersects the characteristics of the four folders. Each intersection determines a dot which symbolizes a comparator, whose color (black or white) represents the sign of output voltage delivered by

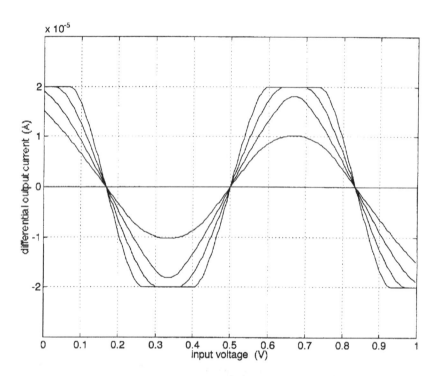

Fig. 8.22 The shape of the differential voltage collected across the symmetrical load of the circuit shown in Fig. 8.21 is controlled by the overlapping of adjacent differential pairs.

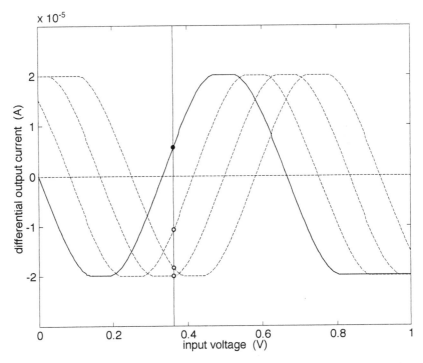

Fig. 8.23 To recover the information contained in the magnitude of the residue of the first generation folders, several shifted folders are put to use to get the same data from the occurrence in time of the zero crossings.

the folder. The fine code is now traceable from the data displayed by the four latches, rather than from the magnitude of the residues. The code is derived from the names of adjacent comparators whose outputs are opposite. Note that these lie necessarily in the linear regions of the residues.

The most striking difference with respect to flash converters is that the number of latches is substantially reduced. In a converter that uses K folders, each M folds wide, only K latches are needed. A flash converter with the same resolution would require $K.M$ latches. In other words, the folding principle divides the number of comparators by the number of folds. How many folds can be implemented without jeopardizing accuracy? Eight seems to be a good compromise, because the magnitude of the folds tends to decrease too rapidly when the currents delivered by the differential pairs overlap excessively. Higher resolutions require thus more folders, but this could lead to more power consumption. Fortunately, one may reach the same goal without overly increasing the power consumption, using resistive dividers between the output terminals of every consecutive folder. The idea,

216 | Fast A to D converters

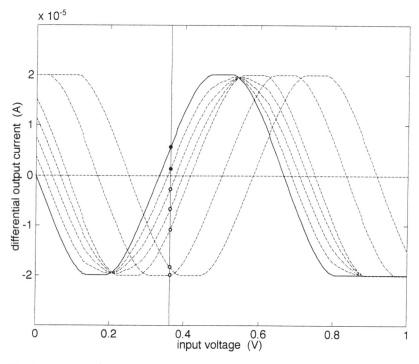

Fig. 8.24 Interpolation offers a means to increase the resolution without the need to add more folders.

called *interpolation*, is shown in Fig. 8.24 by the three additional curves located between the output of the first and second folders. Despite the fact that the interpolated characteristics differ slightly from the original, all the new zero crossings are nicely distributed between those of the original folder. The states of the latches connected at the output of the interpolating dividers now define 4 increments. The hardware requirements however are rather modest: resistive dividers and a few more latches.

The interpolation technique is used in the 8-bit 55 MS/s converter shown in Fig. 8.25 (van de Grift *et al.* 1987). The converter consists of four parallel folding circuits, each 8 folds wide. Interpolation resistors add 2 bits to the 6 bits resolved already by the four folders. Shifting the reference is done by simply connecting each folder to a set of 16 interleaved references derived from 64 taps along a single resistive divider. The coarse code bits are derived from the fold-number, and the fine bits from the 16 comparators. The fold-number, which is the coarse code of the input, can be derived from a separate coarse flash converter.

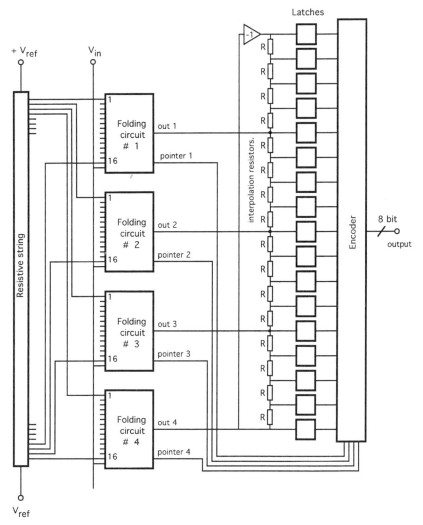

Fig. 8.25 Schematic of an 8-bit folding converter that makes use of four folders, each eight folds wide, plus interpolation.

8.3.3 Comparative study of folding versus flash converters

Figure 8.26 compares the performance of flash and folding converters with those of other fast converters, such as pipelined, subranging and Delta–Sigma. The data refer to publications from 1990 until mid-1999.

BICMOS has been advocated in the past for implementing folding converters (Flynn and Shealan 1998; Vorenkamp and Roovers 1997) as it

218 | Fast A to D converters

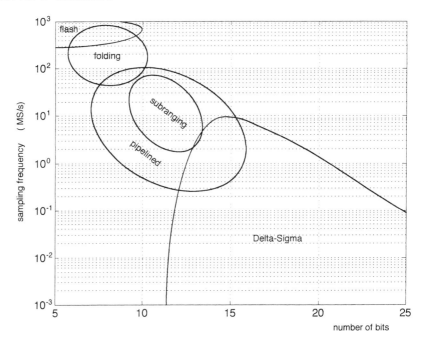

Fig. 8.26 Comparison of the speed/resolution performances of flash, folding, pipelined, subranging and Delta–Sigma converters.

takes advantage of the superior speed performance of analog signal processing hardware. However, CMOS is receiving increasing attention (Venes and van de Plassche 1996; Nauta and Venes 1995). Whichever technology, the sampling rates are not truly competitive with those of pure flash converters. Whether S.H. front-ends are essential parts or not remains an open question. The 400 MS/s converter described by Flynn and Shealan (1998) does not use an S.H, but its resolution is only 6-bits. The 12-bit, 60 MS/s cascaded folding and interpolating converter described by Vorenkamp and Roovers (1997) uses track and hold circuits. Area and power consumption are smaller than with flash converters, since less comparators are needed. Practical power consumption ranges from 80 mW for the 8-bit converter of Venes and van de Plassche (1996), to 200 mW for the 6-bit converter described by Flynn and Shealan (1998), and 300 mW for the 12-bit converter described by Vorenkamp and Roovers (1997).

Smaller numbers of comparators is not the only difference. In flash converters, every comparator comprises a differential pre-amplifier. The size of this amplifier is determined by the maximum tolerable offset. In folding architectures, the differential amplifier is part of the folder. Thus, the incremental steps that must be discriminated are determined by the number of folds. Hence, somewhat larger offsets can be tolerated. These affect the *INL*, rather than the *DNL*, for the interpolators always divide necessarily the space between adjacent folds into almost equal steps.

References

Chapter 1

Boser, B., Karlman, Marti and Wooley, B. (June 1988). Simulating and testing oversampled A to D converters. *IEEE Trans. on computer aided design*, **7**, (6), 668–74.

Burr-Brown Corp. (1987). *The handbook of linear IC applications*, 221–2. Tucson, Arizona.

Doernberg, J., Lee, H.S. and Hodges, D. (Dec. 1984). Full-Speed testing of A/D converters. *IEEE JSSC*, **SC-19**, (6), 820–7.

Ginetti, B. and Jespers, P. (1991). Reliability of code density test for high resolution ADC's. *Electronic letters*, **27**, 2231–3.

Harris, F.J. (1978). On the use of windows for harmonic analysis with the Discrete Fourier Transform. *Proc. IEEE*, **66**, (1), 51–83.

Hoeschele, D.F. (1994). *Analog-to-Digital and Digital-to-Analog conversion techniques*, p. 397. John Wiley, New York.

Razavi, B. (1995). *Principles of data conversion system design*, p. 256. IEEE Press Marketing, Piscataway, N.J., U.S.A.

Chapter 2

Bastiaansen, C.A.A., Groeneveld, D.W., Schouwenaars, H.J. and Termeer, H.A.H. (July 1991). A 10-b 40-MHz 0.8 μm CMOS current-output D/A converter. *IEEE JSSC*, **26**, (7), 917–21.

Bult, Kl. and Geelen, G.J.G.M. (Dec. 1992). An inherently linear and compact MOST-Only current division technique. *IEEE JSSC*, **27**, (12), 1730–5.

Burr Brown (1986). 16 bit resistive DAC with trim. Product Data Book. Tuczon, AZ.

Chi-Huang, L. and Bult, Kl. (Dec. 98). A 10-b, 500-Msample/s CMOS DAC in 0.6 mm^2. *IEEE JSSC*, **33**, (12), 1948–58.

Cremonesi, A., Maloberti, F. and Polito, G. (June 1989). A 100-MHz CMOS DAC for Video-Graphic Systems. *IEEE JSSC*, **24**, (3), 635–9.

Fournier, J.M. and Senn, P. (1990). A 130 MHz 8 bit CMOS video DAC for HDTV applications. *16th. European Solid State Circuits Conference Proc.*, Grenoble, France, 77–80.

Gray, P.R. and Meyer, R.G. (1993). *Analysis and design of analog integrated circuits* (3rd edn), p. 139. John Wiley and Sons, New York.

Guy, T.S. and Trythall, L.M. (1982). A 16 bit monolithic bipolar DAC. *1982 IEEE International Solid-State Circuits Conference*, 88–9.

Hammerschiemd, C.M. and Quiuting, H. (August 1998). Design and implementation of an untrimmed MOSFET-only 10-bit A/D converter with −79 dB THD. *IEEE JSSC*, **33**, (8), 1148–57.

Kakamura, Y., Takahiro, M., Atsushi, M., Harafusa, K. and Nobuharu, Y. (April 1991). A 10-b 70-MS/s CMOS D/A converter. *IEEE JSSC*, **26**, (4), 637–42.

Kamath, B.Y., Meyer, R.G. and Gray, P.R. (Dec. 1974). Relationship between frequency response and settling time of operational amplifiers. *IEEE JSSC*, **SC-9**, 347–52.

Lakshmikumar, K.R., Hadaway, R.A. and Copeland, M.A. (Dec. 1986). Characterization and modeling of mismatch in MOS transistors for precision analog design. *IEEE JSSC*, **SC-21**, (6), 1057–66.

Maio, K., Hotta, M., Yokozawa, N., Nagata, M. and Kaneko, K. (1981). An untrimmed DAC with 14 bit resolution. *1981 IEEE International Solid-State Circuits Conference*, 24–5.

Mc. Creary, J.L. (Dec. 1981). Matching properties and voltage and temperature dependence of MOS capacitors. *IEEE JSSC*, **SC-16**, (6), 608–16.

Pelgrom, M.J.M., Duinmaijer, A.C.J. and Welbers, A.P.G. (Oct. 1989). Matching Properties of MOS Transistors. *IEEE JSSC*, **24**, (5), 1433–40.

Shyu, J.B., Temes, G.C. and Yao, K. (Dec. 1982). Random errors in MOS capacitors. *IEEE JSSC*, **SC-17**, 1070–6.

Shyu, J.B., Temes, G.C. and Krummenacher, F. (Dec. 1984). Random error effects in matched MOS capacitors and current sources. *IEEE JSSC*, **SC-19**, (6), 948–55.

Takahiro, M., Yasuyuki, N., Masao, N., Sotoju, A., Yoichi, A. and Yasutaka, H. (June 1986). An 80 MHz 8 bit CMOS D/A converter. *IEEE JSSC*, **SC-21**, (6), 983–8.

Tsividis, Y.P. (1987). *Operation and modeling of the MOS transistor*. Mc Graw Hill, New York.

van den Elshout (1994). Analysis of an M-2M current divider for a mixed signal sea of gates array. *Unpublished D. Phil. thesis*. University of Twente, The Netherlands.

Wittmann, R., Schardein, W., Hosticka, B.J., Burbach, G. and Arndt, J. (August 1995). Trimless high precision ratioed resistors in D/A and A/D converters. *IEEE JSSC*, **30**, (8), 935–9.

Yee, Y.S., Termann, L.M. and Heller, L.G. (August 1979). A two-stage weighted capacitor network for D/A–A/D conversion. *IEEE JSSC*, **SC-14**, (4), 778–81.

Chapter 3

Daubert, S.J., Vallancourt, D. and Tsividis, Y.P. (Dec. 1988). Current copier cells. *Electronics Letters*, **24**, (25), 1560–2.

Fotouhi, B. and Hodges, D.A. (Dec. 1979). High-Resolution A/D Conversion in MOS/LSI. *IEEE JSSC*, **SC-14**, (6), 920–6.

Macq, D. and Jespers, P. (April 1993). Charge injection in current copier cells. *Electronics Letters*, **29**, (9), 780–1.

Schoeff, J.A. (Dec. 1979). An inherently monotonic 12 bit DAC. *IEEE JSSC*, **SC-14**, 904–11.

Schouwenaars, H.J., Dijkmans, E.C., Kup, B.M.J. and Van Thuijl, E.J.M. (June 1986). A monolithic dual 16-bit D/A converter. *IEEE JSSC*, **SC-21**, (3), 424–9.

Schouwenaars, H.J., Wouters, D., Groeneveld, J. and Termeer, H.A.H. (Dec. 1988). A low-power stereo 16-bit CMOS D/A converter for digital audio. *IEEE JSSC*, **23**, (6), 1290–7.

Tuthill, M. (1980). A 16 bit Monolithic CMOS D/A Converter. *6th. European Solid State Circuits Conference Proc.*, Grenoble, France, 353–5.

Van de Plassche, R.J. (Dec. 1976). Dynamic element matching for high-accuracy monolithic D/A converters. *IEEE JSSC*, **SC-11**, (6), 795–800.

Van de Plassche, R.J. and Goedhart, D. (June 1979). A monolithic 14 bit D/A converter. *IEEE JSSC*, **SC-14**, (3), 552–6.

Vittoz, E.A. and Wegmann, G. (1990). Dynamic current mirrors (Chapter 7). In *Analogue IC design: the current-mode approach*. (1st edn) (ed. C. Toumazou, F.J. Lidgey and D.G. Haig), 297–324. P. Peregrinus Ltd., London.

Wegmann, G., Vittoz, E.A. and Rahali, F. (Dec. 1987). Charge injection in MOS switches. *IEEE JSSC*, **SC-22**, (6), 1091–7.

Wouters, D., Groeneveld, J., Schouwenaars, H.J., Termeer, H.A.J. and Bastiaansen, C.A.A. (Dec. 1989). A self-calibration technique for monolithic high-resolution D/A converters. *IEEE JSSC*, **24**, (6), 1517–22.

Chapter 4

CCITT (1992). Transmission performance characteristics of pulse code modulation. Recommendation G.712 (09/92). International Telecommunication Union.

Creary, J.L. and Gray, P.R. (Dec. 1975). All MOS Charge Redistribution Analog to Digital Conversion Techniques, Part 1. *IEEE JSSC*, **SC-10**, (6), 371–9.

Lee, H-S., Hodges, D.A. and Gray, P.R. (Dec. 1984). A self-calibration 15 bit CMOS A/D converter. *IEEE JSSC*, **SC-19**, (6), 813–49.

Ohri, K.B. and Callahan, M.J. (Feb. 1979). Integrated PCM codec. *IEEE JSSC*, **SC-14**, (1), 38–48.

Chapter 5

Chen, C.C. and Wu, C-Y. (Jan. 1998). Design techniques for 1.5-V low-power CMOS current-mode cyclic analog-to-digital converters. *IEEE Trans. on circuits and systems*, II: analog and digital signal processing, **45**, (1), 28–40.

Degrauwe, M., Vittoz, E. and Verbouwhede, I. (June 1985). A micropower CMOS-instrumentation amplifier. *IEEE JSSC*, **SC-20**, 805–7.

Deval, Ph., Robert, J. and Declercq, M.J. (1991). A 14 bit CMOS A/D converter based on dynamic current memories. *Custom integrated circuits conference*, 24.2.1–24.2.4.

Gani, J. and Gray, P.R. (August 1990). A 1-bit/cycle algorithmic analog-to-digital converter without high-precision comparators. *Memorandum n° UCB/ERL M90/69* Electronics Research Laboratory, U.C. Berkeley.

Ginetti, B. (March 1992). CMOS RSD cyclic A-to-D converters. Doctoral thesis. UniversitéCatholique de Louvain.

Ginetti, B., Jespers, P.G.A. and Vandemeulebroecke, A. (July 1992). A CMOS 13-b cyclic RSD A/D converter. *IEEE JSSC*, **27**, (7), 957–65.

Grisoni, L., Heubi, A., Balsinger, P. and Pellandini, F. (Sept. 1997). Micro power 14-bit A/D converter : 45 μW at +/– 1.25 V and 16 kSamples/s. *Int. symp. on I.C. technology, systems & applications*. Singapore, 39–42.

Hwang, K. (1979). *Computer arithmetic—principles, architecture and design*. J. Wiley, New York.

Li, P.W., Chin, M.J. and Gray, P. (Dec. 1984). A ratio-independent algorithmic analog-to-digital conversion technique. *IEEE JSSC*, **SC-19**, (6), 828–36.

Macq, D. and Jespers, P.G.A. (August 1994). A 10-bit pipelined switched-current A/D converter. *IEEE JSSC*, **29**, (8), 967–71.

McCharles, R.H., Saletore, V.A., Black, W.C. and Hodges, D.A. (1977). An algorithmic analog to digital converter. *IEEE Int. Solid State Circuits Conference Digest of Tech. Papers*.

Nairn, D.G. and Salama, A.T. (March 1990). A ratio-independent analog-to-digital converter combining current mode and dynamic techniques. *IEEE Trans. CAS*, **37**, (3), 319–25.

Ohara, H., Ngo, H.X., Armstrong, M.J., Chowdhury, F.R., Gray, P. (Dec. 1987). A CMOS programmable self-calibrating 13-bit eight-channel data acquisition peripheral. *IEEE JSSC*, **SC-22**, (6), 930–8.

Shih, C.C. and Gray, P.R. (August 1986). Reference refreshing cyclic analog-to-digital and digital-to-analog converters. *IEEE JSSC*, **SC 21**, (4), 544–54.

Shih, C.C., Li, P.W. and Gray, P. (Oct. 1983). Ratio independent Cyclic A/D and D/A conversion using a recirculating reference approach. *IEEE Trans CAS*, **30**, (10), 772–4.

Wang, Z. (June 1991). Design methodology of CMOS algorithmic current A/D converters in view of transistor mismatches. *IEEE Trans. on circuits and systems*, **38**, (6), 660–7.

Chapter 6

Kayanuma, A., Takeda, M., Sugiyame, S., Iga, A., Kobayashi, T. and Asano, K. (1981). An integrated 16 bit A/D converter for PCM audio systems. *1981 IEEE International Solid-State Circuits Conference*, 56–7.

Musa F.H. and Hungtinton R.C. (1976). A CMOS monolithic 3 1/2-digit A/D converter. *1976 IEEE International Solid State Circuits Conference*, 144–5.

Pelgrom, M.J.M. and Roorda, M. (1988). An algorithmic 15 bit CMOS D-A Converter. *1988 IEEE International Solid-State Circuits Conference*, 198–9.

Chapter 7

Ardalan, S.H. and Paulos, J.J. (June 1987). An analysis of nonlinear behavior in delta-sigma modulators. *IEEE Trans. CAS*, **34**, (6), 593–603.

Baird, R.T. and Fiez, T.S. (Jan. 1994). Stability analysis of high-order delta-sigma modulation for ADC's. *IEEE Trans. CAS, II Analog and Digital Signal Proc.*, **41**, (1), 59–62.

Baird, R.T. and Fiez, T.S. (March 1996). A low oversampling ratio 14-b 500-kHz $\Delta\Sigma$ ADC with a self-calibrated multibit DAC. *IEEE JSSC*, **31**, (3), 312–20.

Bazarjani S. and Snelgrove, M. (May 1998). A 160-MHz fourth-order double-sampled SC bandpass sigma–delta modulator. *IEEE Trans. CAS, II analog and digital signal processing*, **45**, (6), 547–55.

Benabes, P., Gauthier, A. and Kielbasa, R. (1996). A multistage closed-loop sigma–delta modulator (MSCL). *Analog integrated circuits and signal processing*, **11**, 195–204. Kluwer, Boston.

Boser, B.E. and Wooley, B.A. (Dec. 1988). The design of sigma–delta modulation analog-to-digital converters. *IEEE JSSC*, **23**, (6), 1298–1308.

Brooks, T.L., Robertson, D.H., Kelly, D.F., Del Muro, A. and Harston, St. W. (Feb. 1997). A 16b $\Sigma\Delta$ pipeline ADC with 2.5 MHz output data rate. *ISSCC 1997 Digest of Techn. Papers*, 208–9.

Candy, J.C. and Temes, G.C. (1992). *Oversampling delta-sigma data converters, theory, design and simulation*, 499. IEEE Press, Piscataway, N.J.

Chao, K.C.H., Nadeem, S., Lee, W.L. and Sodini, Ch. (March 1990). A higher order topology for interpolative modulators for oversampling A/D converters. *IEEE Trans. C.A.S.*, **37**, (3), 309–18.

Chuang, S., Liu, H., Sculley, T.L. and Bamberger, R.H. (May 1998). Design and implementation of bandpass delta–sigma modulators using half-delay integrators. *IEEE Trans. CAS II analog and digital signal processing*, **45**, (6), 535–45.

Crochiere, R.E. and Rabiner, L.R. (March 1981). Interpolation and decimation of digital signals—a tutorial review. *IEEE Proc.*, **69**, 300–31.

Daubert, S.J. and Vallancourt, D. (1991). A transistor-only current-mode $\Sigma\Delta$ modulator. *IEEE Custom Integrated Circuits Conference*, 24.3.1–24.3.4.

Frank Op 't Eynde and Sansen, W. (1993). *Analog interfaces for digital signal processing systems*, p. 251. Kluwer Academic Press, Boston/Dordrecht/London.

Goedhart, D., van de Plassche, R.J. and Stikvoort, E.F. (1982). Digital-to-analog converstion in playing a Compact Disc. *Philips Technical Review*, **40**, (6), 174–9.

Huijsing, J.H., van de Plassche, R.J. and Sansen, W. (eds) (1993). *Analog circuit design. Analog to digital converters*. Kluwer Academic Publishers. Dordrecht.

Jantzi, St. A., Snelgrove, W.M. and Ferguson, P.F. (March 1993). A fourth-order bandpass $\Sigma\Delta$ modulator. *IEEE JSSC*, **28**, (3), 282–91.

Jantzi, St. A., Martin, K. and Sedra, A. (Dec. 1997). Quadrature bandpass $\Sigma\Delta$ modulation for digital radio. *IEEE JSSC*, **32**, (12), 1935–50.

Leslie, T.C. and Singh, B. (June 1992). Sigma–delta modulators with multibit quantising elements and single-bit feedback. *IEE Proceedings-G*, **139**, (3), 356–62.

Leung, B.H. (Oct. 1991). Design methodology of decimation filters for oversampled ADC based on quadratic programming. *IEEE Trans. CAS*, **38**, (10), 1121–32.

Leung, B.H. and Sutarje, S. (Jan. 1992). Multibit Σ-Δ A/D converter incorporating a novel class of dynamic element matching techniques. *IEEE Trans. CAS. II Analog and Digital Signal Proc.*, **39**, (1), 35–51.

Ma, Z.P. and Leung, B. (August 1992). Polyphase IIR decimation filter design for oversampled A/D converters with approximately linear phase. *IEEE Trans. CAS II, Analog and Digital Signal Proc.*, **39**, (8), 497–505.

Magrath, A.J. and Sandler, M.B. (May 1995). Efficient dithering of sigma–delta modulators with adaptative bit flipping. *Electronic letters*, **31**, (11), 846–7.

Moeneclay, N. and Kaiser, A. (July 1997). Design techniques for high resolution current-mode sigma–delta modulators. *IEEE JSSC*, **32**, (7), 953–8.

Norsworthy, St.R., Schreier, R. and Temes, G.C. (1997). *Delta–sigma data converters, theory, design and simulation.* IEEE Press, Piscataway, N.J.

Nys, O. and Henderson, R. (1996). A monolythic 19-bit 800 Hz low power multi-bit sigma–delta CMOS ADC using data weighting averaging. *Convention record of ESSCIRC,* **96,** 252–5.

Peluso, V., Vancorenland, P., Marques, A. and Sansen, W. (Feb. 1998). A 900 mV 40 µW switched Opamp. $\Delta\Sigma$ modulator with 77 dB dynamic range. *ISSCC 1998 Digest of Technical Papers,* 68–9.

Schouwenaars, H., Groeneveld, W., Bastiaansen, C. and Termeer, H. (April 1992). Continuous calibration noise shaping D/A conversion. In *Advances in Analog Circuit Design,* 243–65. (Huijsing, J.H., van de Plassche, R.J. and Sansen, W. (eds) Kluwer Academic Publishers: Dordrecht.

Shreier, R. (August 1994). On the use of chaos to reduce idle-channel tones in delta–sigma modulators. *IEEE Trans. CAS, I Theory and Applications,* **41,** (8), 539–47.

Schreier, R. and Snelgrove, M. (Nov. 1989). Bandpass sigma–delta modulation. *Electronic letters,* **25/23,** 1560–1.

Sherstinsky, A. and Sodini, C.G. (Sept. 1990). A programmable demodulator for oversampled analog-to-digital modulators. *IEEE Trans. C.A.S.,* **37,** (9), 1092–103.

Stikvoort, E.F. (Oct. 1988). Some remarks on the stability and performances of the noise shaper or sigma–delta modulator. *IEEE Trans. Communications,* **36,** (10), 1157–62.

Uchimura, K., Hayashi, T., Kimura, T. and Iwata, A. (Dec. 1988). Oversampling A-to-D and D-to-A converters with multistage noise shaping modulators. *IEEE Trans. Acoust, Speech and Signal Proc.* **36,** (12), 1899–905.

van de Plassche, R.J. (July 1978). A Sigma–Delta modulator as an A/D converter. *IEEE Trans. CAS,* **25,** (7), 510–14.

van de Plassche, R.J. (1994). *Analog-to-digital and digital-to-analog converters.* p. 548. Kluwer Academic Publishers; Dordrecht.

Chapter 8

Akazawa, Y., Iawata, A., Wakimoto, T., Nakamura, H. and Ikawa, H. (Feb. 1987). A 400 MSPS 8 bit Flash AD Conversion LSI. *1987 ISSC digest of technical papers,* 98–9.

Brooks, T.L., Robertson, D.H., Kelly, D.F., Del Muro, A. and Harston, St.W. (Dec. 1997). A cascaded sigma–delta pipeline A/D converter with 1.25 MHz signal bandwidth and 86 dB SNR. *IEEE JSSC,* **32,** (12), 1896–906.

Bult, Kl. and Buchwald, A., (Dec. 1997). An embedded 240-mW 10-bit 50-MS/s CMOS ADC in 1 mm^2. *IEEE JSSC,* **21,** (12), 1887–95.

Byunghak, Th. and Gray, P.R. (March 1995). A 10-b, 20 Msamples/s, 35 mW pipeline A/D converter. *IEEE JSSC,* **30,** (3), 166–72.

Cline, D.W. and Gray, P.R. (March 1996). A power optimized 13-b 5 Msamples/s pipelined analog-to-digital converter in 1.2 µm CMOS. *IEEE JSSC,* **31,** (3), 294–303.

Colleran, W.T. and Abidi, A.A. (Dec. 1993). A 10-b, 75-MHz two-stage pipelined bipolar A/D converter. *IEEE JSSC,* **28,** (12), 1187–99.

Dingwall, A.G.F. (Feb. 1985). An 8 MHz 8b CMOS subranging ADC. *ISSCC Digest of Techn. Papers*, 72–3.
Flynn, M.P. and Shealan, B. (Dec. 1998). A 400-Msample/s, 6-b CMOS folding and interpolating ADC. *IEEE JSSC*, **33**, (12), 1932–8.
Ginetti, B. (1992). CMOS RSD cyclic A-to-D converters. *Doctoral thesis*. (chapter 4). Universté Catholique de Louvain.
Hadidi, K. and Temes, G.C. (August 1992). Error analysis in pipeline A/D converters and its applications. *IEEE Trans. CAS, II analog and digital signal processing*, **39**, (8), 506–15.
Hosotani, S., Miki, T., Maeda, A. and Yazawa, N. (Feb. 1990). An 8-bit 20MS/s CMOS A/D converter with 50-mW power consumption. *IEEE JSSC*, **25**, (1), 167–72.
Huijsing, J.H., van de Plassche, R.T.J. and Sansen, W. (1993). *Analog circuit design. High speed folding ADC's*, 163–83. Kluwer Academic Publishers, Dordrecht.
Karanicolas, A.N. and Lee, H-S. (Dec. 1993). A 15-b 1-Msample/s digitally self-calibrated pipelined ADC. *IEEE JSSC*, **28**, (12), 1207–15.
Kattmann, K. and Barrow, J. (Feb. 1991). A technique for reducing differential non linearity errors in flash A/D converters. *ISSCC Digest technical Papers*, (S.F., Ca), 170–1.
Kazuya S., Nishida Y. and Nakadai, N. (Dec. 1993). A 10-b 100 Msample/s pipelined subranging BICMOS ADC. *IEEE JSSC*, **28**, (12), 1180–6.
Kim, K.Y., Kusanayagi, N. and Abidi, A.A. (March 1997). A 10-b, 100-MS/s CMOS A/D converter. *IEEE JSSC*, **32**, (3), 302–11.
Krishnaswamy, N., Fetterman, H.S., Anidjar, J., Lewis, St.H. and Renninger, R.G. (March 1997). A 250-mW, 8-b, 52 Msamples/s parallel-pipelined A/D converter with reduced number of amplifiers. *IEEE JSSC*, **32**, (3), 312–20.
Kusumoto, K., Matsuzawa, A. and Murata, K. (Dec. 1993). A 10-b 20-MHz 30-mW pipelined interpolating CMOS ADC. *IEEE JSSC*, **28**, (12), 1200–6.
Kwak, S-U., Song, B-S. and Bacrania, K. (Dec. 1997). A 15-b, 5-Msamples/s low-spurious CMOS ADC. *IEEE JSSC*, **32**, (12), 1866–75.
Lee, S-H. and Song, B-S. (Dec. 1992). Digital domain calibration of multistep analog-to-digital converters. *IEEE JSSC*, **27**, (12), 1679–88.
Lewis, S.H. and Gray, P.R. (Dec. 1987). A pipelined 5 Msample/s 9-bit analog-to-digital converter. *IEEE JSSC*, **SC-22**, (6), 954–61.
Lewis, S.H. (August 1992). Optimizing the stage resolution in pipelined, multistage, analog-to-digital converters for video-rate applications. *IEEE Trans. CAS, II analog and digital signal processing*, **39**, (8), 516–23.
Lewis, S.H., Fetterman, H.S., Gross, G.F., Ramachandran, R. and Viswanathan, T.R. (March 1992). A 10-b 20 Msample/s analog to digital converter. *IEEE JSSC*, **27**, (3), 351–8.
Lin, Y.M., Kim, B., Gray, P.R. (April 1991). A 13-b 2.5-MHz self-calibrated pipelined A/D converter in 3-µm CMOS. *IEEE JSSC*, **26**, (4), 628–36.
Macq, D. and Jespers, P.G.A. (August 1994). A 10-bit pipelined switched current A/D converter. *IEEE JSSC*, **29**, (8), 967–71.
Mangelsdorf, C.W. (1990). A 400-MHz input flash converter with error correction. *IEEE JSSC*, **25**, (1), 184–91.
Mayes, M.M. and Chin, S.W. (Dec. 1996). A 200 mW, 1 Msample/s, 16-b pipelined A/D converter with on-chip 32-b microcontroller. *IEEE JSSC*, **31**, (12), 1862–72.

References

Nauta, B. and Venes, A.G.W. (Dec. 1995). A 70 MS/s 110 mW 8-b CMOS folding and interpolating A/D converter. *IEEE JSSC*, **30**, (12), 1302–8.

Peetz, B., Hamilton, B.D. and Kang, J. (Dec. 1986). An 8-bit 250 Megasample per second analog-to-digital converter: operation without a sample and hold. *IEEE JSSC*, **SC-21**, (6), 997–1002.

Pelgrom, M.J.M., Rens, A.C., Vertregt, M. and Dijkstra, M.B. (August 1994). A 25-MS/s 8-bit CMOS A/D converter for embedded applications. *IEEE JSSC*, **29**, (8), 879–86.

Peterson, J. (Dec. 1979). A monolithic video A/D converter. *IEEE JSSC*, **SC-14**, 932–7.

Portmann, C.L. and Meng, T.H.Y. (August 1996). Power-efficient metastability error reduction in CMOS flash A/D converters. *IEEE JSSC*, **31**, (8), 1132–40.

Schmitz, A.E., Walden, R.H., Montes, M., DuChesne, M. and Stevens, E. (1988). A 4 bit, 1 GHz analog-to-digital technology using focused ion beam implants and sub-half micrometer CMOS/SOS. *1988 Symposium on VLSI technology, Digest of Techn. Papers*, 67–8.

Shinagawa, M. Akazawa, Y. and Wakimoto, T. (Feb. 1990). Jitter analysis of high-speed sampling systems. *IEEE JSSC*, **25**, (1), 220–4.

Shu, T.-H., Song, B-S. and Bacrania, K. (April 1995). A 13-b 10-Msamples/s ADC digitally calibrated with oversampling delta-sigma converter. *IEEE JSSC*, **30**, (4), 443–52.

Song, B.S., Lee, S.H. and Tompsett, M.F. (Dec. 1990). A 10 bit 15 MHz CMOS recycling two-step A-D converter. *IEEE JSSC*, **25**, (6), 1328–38.

van de Grift, R.E.J., Rutten, I.W.J.M. and van der Veen, M. (Dec. 1987). An 8-bit video ADC incorporating folding and interpolation techniques. *IEEE JSSC*, **22**, (6), 944–53.

van de Plassche, R.J. and Baltus, P. (Dec. 1988). An 8-bit 100-MHz full Nyquist analog-to-digital converter. *IEEE JSSC*, **23**, (6), 1334–44.

van Valburg, J. and van de Plassche, R.J. (1993). High speed folding ADC's. *Analog circuit design*, 163–83. Kluwer, Dordrecht.

Venes, A.G. and van de Plassche, R.J. (Dec. 1996). An 80-MHz, 90-mW, 8-b CMOS folding A/D converter with distributed track-&-hold preprocessing. *IEEE JSSC*, **31**, (12), 1848–53.

Vorenkamp, P. and Roovers, R. (Dec. 1997). A 12-b, 60 Msample/s cascaded folding and interpolating ADC. *IEEE JSSC*, **32**, (12), 1876–86.

Wakimoto, T., Akazawa, Y. and Konaka, S. (Dec. 1988). Si bipolar 2-GHz 6-bit flash A/D conversion LSI. *IEEE JSSC*, **23**, (6), 1345–50.

Yotsuyanagi, M., Toshiyuki, E. and Kazumi, H. (March 1993). A 10-b 50-MHz pipelined CMOS A/D converter with S/H. *IEEE JSSC*, **28**, (3), 292–300.

Yu, P.C. and Lee, H-S. (Dec. 1996). A 2.5 V, 12-bit, 5-Msample/s pipelined CMOS ADC. *IEEE JSSC*, **32**, (12), 1854–61.

Appendix

Introduction: why simulations?

Computer simulations offer a sensible way to understand the operation of D to A and A to D converters more deeply. A clear distinction must be made however between transistor- and system-level simulations. Transistor-level simulations analyze the detailed behavior of circuits. Running them is very costly in terms of computer time because the calculations rely upon very large numbers of time samples, each resolving a large number of transistors. System-level simulations, on the contrary, globalize the behavior and therefore do not suffer from the same drawback.

The MATLAB program available on the website is intended for system-level simulation. The principles underlying the associated converter simulation toolbox are explained below. The text concludes with a few examples to illustrate how to carry out converter 'software experiments'.

www.oup.co.uk/best.textbooks/engineering/jespers

1. 'Analog' and 'digital' representations

The interpretation of simulation results rests mostly on graphs, like linear and logscaled plots. Since converters combine analog and digital data, the representation of the latter is entrusted to their analog counterparts. Thus, binary coded words are represented by 2^N discrete DC levels as illustrated by the quantization function **qtz**(in,N), whose input 'in' is 'analog' (continuous) and output 'digital' (discrete). Consider the MATLAB program below:

```
clear
N = 3;                   % number of bits
in = -1.5:.01 : 1.5;     % input vector
out = qtz(in,N);         % output vector
stairs(in,out); grid     % plot
```

The plot of Fig. A.1 represents the transfer characteristic of a 3-bit quantizer versus the continuous analog input with eight discrete plateaus illustrating the set of all the output words of the quantizer (unless specified, dynamic ranges are normalized between minus and plus one).

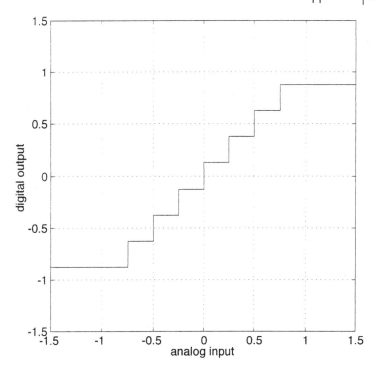

Fig. A.1.

2. The input data

Testing converters requires appropriate input data (generally a linear ramp or a pure sine wave). The linear ramp is used in order to assess the DC characteristics, like the transfer function and INL–DNL plots whereas a pure sine wave is used to evaluate the AC performances, like in the code density test or spectral signature.

In the next paragraphs we consider how to input 'analog' and 'digital' data to A to D and D to A converters.

2.1 Input data for A to D converters

The input data required to test A to D converters are generally linear input ramps and sinusoidal signals. A linear ramp can be obtained by means of the MATLAB command

in = **linspace**(lower bound, upper bound, number of samples);

'Analog' sine waves can be invoked similarly by means of MATLAB instructions. We recommend however the function **sinput**(St,Np,P,D), which is intended especially for spectral signature evaluation. **Sinput** generates a matrix consisting of St columns, each containing a sine wave whose magnitude is normalized to one. Each column contains the same number 'Np' of entire sine wave periods but with random phases. The number of rows represents the number of input samples Ns accordingly to the equation below:

$$Ns = 2^{\wedge}(P) + D$$

'D' adds D more samples to the set already defined by 'P' to allow eventual transients to die out before spectral analysis. It is strongly recommended to make use of this variable when systems with state variables are being considered, like Delta–Sigma converters. Making D equal to 50 or 100 is generally enough to avoid the deterioration of the **ffts**. This delay is transparent for it is erased automatically once spectra are being evaluated. The number of samples is then a power of two. When not specified, D is automatically equal to zero.

The example below shows a set of instructions that defines a matrix Y consisting of 5 columns, each 1024 samples long representing 2 periods of the same sine wave with random phases (Fig. A.2):

```
clear
St = 5; Np = 2; P = 10;
Y = sinput(St,Np,P);
Ns = 2^(P); time = 1: Ns;
plot(time, Y)
```

2.2 Input data for D to A converters

The simulation of D to A converters requires a little more caution since D to A converters are controlled by coded input words. A look-up table is used for this purpose.

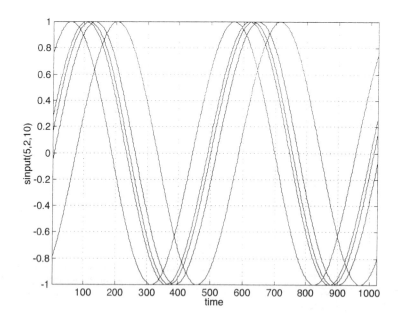

Fig. A.2.

Look-up tables can be implemented by combining the instructions **binm**(N) or **trm**(N) with **binw**(Typ,N,sigm). The instructions **binm** and **trm** generate matrices or tables representing the ordered sets of all possible binary codes that can be represented by *N*-bit words. The **binm** instruction applies to single quadrant *binary* coded words whereas **trm** lists *redundant signed digit* (RSD) words (the latter is used widely in this toolbox for it removes automatically DC pedestals, easing **fft** calculations). The instruction **binw**(Typ,N,sigm) which is described further, generates a column vector *N* rows long that represents the weights in descending order of any binary coded D to A converter described by the variables *Typ*, *N* and *sigm*. When **binm** or **trm** is multiplied by the column vector **binw**, the results is the look-up vector of the D to A converter.

In the example below the binary and RSD look-up tables w1 and w2 of a perfect 3-bit D to A scaled converter are generated.

```
clear
N = 3;
w1 = binm(N)*binw(2,N,0);
w2 = trm(N)*binw(2,N,0);
w = [ w1 w2]
```

The result is:

w =

0	− 0.8750
0.1250	− 0.6250
0.2500	− 0.3750
0.3750	− 0.1250
0.5000	0.1250
0.6250	0.3750
0.7500	0.6250
0.8750	0.8750

In the example below, we apply a 'digital' two periods sine wave consisting of 256 samples to the converter above using the look-up vector **w2**. The 'digital' sine wave is generated first starting from the analog **sinput** instruction, which is followed by a **qtz** instruction that generates the 'analog' counterpart of the digital input sine wave. The next instruction is a linear transform changing the discrete outputs of the quantizer into addresses z, which access the look-up table 'w2'. The 'analog' output of the converter is the result of instruction w2(z):

```
clear
N = 3;
% generation of the look-up table
w2 = trm(N)*binw(2,N,0);
% 'analog' input sine wave
St = 1; Np = 2; P = 8; in = sinput(St,Np,P);
% quantized sine wave
q = qtz(in,N);
% address conversion
z = 2^(N-1)*(q+1) +.5;
```

```
% output
out = w2(z,:);
% graph of the 'analog' input before the quantizer and converter
output
X = 1:2^(P); plot(X,[ in out] )
```

The outcome is illustrated in Fig. A.3.

The same procedure may be used to produce a 'digital' ramp to evaluate the converter static transfer characteristic but this is not very useful since look-up tables are transfers functions by themselves.

3. Available conversion functions

A survey of the available conversion functions is presented below.

3.1 D to A converters

Three types of scalers are considered: *R2R*, *binary unit-elements* and *thermometer* scalers. The unit-elements may either represent unit-capacitors or unit-current source.

The instruction that generates the weight scale of any binary weighted D to A converter is: **binw** (Typ,N,sigm). Typ defines the kind of scaler, 1 for an R2R and 2 for a unit-element binary scaler. N represents the number of bits. Errors affecting the resistors or the unit-elements are controlled by the variable sigm that determines

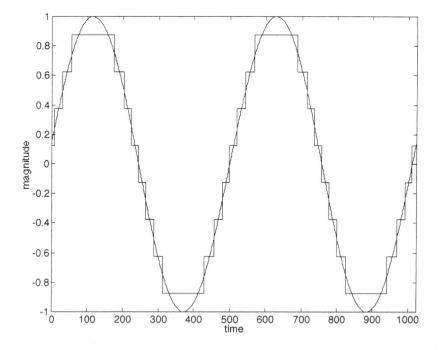

Fig. A.3.

the standard deviation of the Gaussian distribution of the errors. For instance, the instruction:

$z = $ **binw**$(2,8,0.01)$

generates a column vector **z** which represents the binary weights in descending order of an 8-bit unit-elements binary scalar whose unit-elements exhibit errors of 1% standard deviation with respect to their nominal values.

z = 0.5001
     0.2497
     0.1249
     0.0627
     0.0312
     0.0157
     0.0080
     0.0039

It is possible to compare the impact of mismatches affecting R2R and unit-element converters. In the first case, the precision of the resistors must be almost the same as that of the converter to comply with the alleged resolution, whereas in the second, large numbers of unit-elements to implement the MSBs improve the overall accuracy.

Multiplying **binm** or **trm** by **binw** generates the look-up table of any D to A converter described by **binw**. An example is shown below:

```
clear
N = 6; d = 2^(-N);
Typ = 2;                        % unit-elements binary scaled conv
sigm = 0.2;                     % stand dev of every unit-element
Y = trm(N)*binw(Typ,N,sigm);    % two-quadrant look-up table
X = -1+d: 2*d: 1-d;             % input codes
plot(X,Y,'+'); grid; axis ('square')
```

The actual transfer characteristic is shown in Fig. A.4.

Notice that in the above example, the input X is equivalent to the cascading of a **linspace** and a **qtz** instruction.

The instruction **da1**(in,N,Typ,sigm,S,sigmS) has been introduced in order to bypass most of the above instructions. It computes the analog output of any N-bit scaled D to A converter or segment converter whose digital input is given by the vector or matrix 'in'. 'Typ' defines the type of scaler like above and 'sigm' the standard deviation of every scaler element. The two other variables are optional. They are specified when segment converters are considered. The number of segments is equal to 2^s and the standard deviation of the references defined by 'sigmS'. When omitted, S is equal to zero and the number of segments equal to one so that the converter resumes to a simple binary scaled converter.

For instance, the program at the end of Section 2.2, when it is rewritten as shown below, produces the same plot as the one depicted in Fig. A.3. The instruction da1 takes care automatically of te generation of the look-up table, quantization of the sine wave and address conversion.

```
clear
N = 3;
% 'analog' input sine wave
St = 1; Np = 2; P = 8; in = sinput(St,Np,P);
```

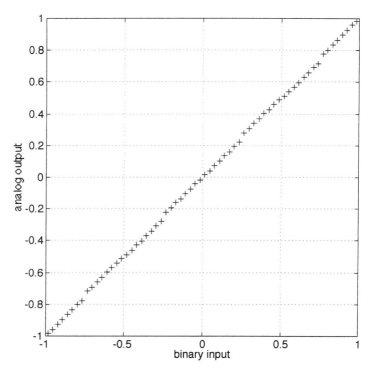

Fig. A.4.

```
% 'digital' output
out = da1 (in,N,2,0);
% graph of the 'analog' input before the quantizer and converter output
X = 1:2^(P); plot (X,[ in out] )
```

The instruction **da2**(in,N,sigm) is intended for N-bit scaled D to A converters whose unit-elements are selected according to a thermometer code instead of binary law. This type of converter exhibits a transfer characteristic without DNL errors in practice. The input is represented by the variable 'in' like above, 'N' being the number of bits and 'sigm' the standard deviation of the unit-elements.

3.2 A to D converters

Various A to D conversion functions are available. They may be divided into three categories, *serial converters* like algorithmic converters, *Delta–Sigma converters* and *multi-step converters*.

Algorithmic converters

The instruction **algor1**(in,N,A,offOpA,offComp) computes the digital output of any N-bit cyclic or algorithmic converter. Impairments can be introduced by means of appropriate choices of the gain A (nominally equal to 2), as well as the Op. Amp. and comparator offsets.

Appendix | 235

The instruction **algor2**(in,N,A,Vmin,Vmax,offOpA) simulates the behavior of RSD cyclic converters. Besides the variables 'A' and 'offOpA', which have the same meanings as above, the comparison levels 'Vmin' and 'Vmax' can be adjusted separately (the nominal values are –0.25 and +0.25 respectively, Vref being always equal to 1). The RSD converter ignores the impact of transition position errors upon the INL and DNL. Reversing the sign of the input after the first cycle compensates moreover the global offset of the transfer characteristic.

Delta–Sigma converters

The three basic noise shapers shown in Fig. 7.13 can be implemented by means of the instruction **dtsgX**(in,N,G,B,sigmAD,sigmDA) where X is equal to 1, 2 or 3, accordingly to the order of the noise shaper. The quantizer consists of an A to D converter followed by a D to A converter whose number of bits is fixed by the variable N. The data outputted by the A to D converter are the actual output data.

Various parameters can be modified in order to evaluate the impact of errors on the overall performances of noise shapers. 'G' defines the feedback loop gain of the accumulators. These are the discrete counterparts of continuous time integrators. When G is smaller than 1, the accumulators are similar to integrators having gains equal to $1/(1-G)$. The resulting 'leakage' from the finite gain of the Op Amps increases the low frequency noise. The variable 'B' controls the symmetrical saturation limits of the integrator located before the quantizer. B offers a means to evaluate the consequences of saturation of the loop filter. The variables 'sigmAD' and 'sigmDA' model the quantizer imperfections. As suggested by the names, the first affects the A to D conversion, the second the D to A. The A to D converter is a flash converter, which consists of a resistive divider ladder followed by a bank of ideal comparators. Impairments are introduced in the string of resistors that form the reference divider by means of the 'sigmAD' variable. The D to A converter is a scaled converter whose unit-elements are controlled by 'sigmDA' in the same way as the **binw** instruction. It is possible to verify that the A to D converter impairments produce unnoticeable effects upon the spectral content of the noise shaper whereas impairments affecting the D to A converter introduce harmonics and increase the noise floor in the baseband. Those effects disappear of course when the quantizer counts only 1 bit.

The duration of the input signals should always be long enough to overcome the artefacts that affect the fft's. A good figure of the variable 'P' defining the sample length $2^{\wedge}(P)$ is 12 to 14 (remember however that the duration of the computations increases exponentially with 'P'). The variable D should be not less than 50 to avoid transients.

Multi-step converters

Multi-step converters are built by means of the instructions **recycl**(in,M,cycl,imp,ext) or **recyclRSD**(in,M,cycl,imp,ext). Each instruction defines a bloc consisting of an M-bit flash converter and an M-bit multiplying D to A converter (MDAC). The input signal 'in' is applied to the flash converter whose output drives the MDAC. The latter is a unit-elements capacitive D to A converter, which outputs the difference between the input and the 'analog' counterpart of coded words delivered by the flash converter. This difference is multiplied by $2^{\wedge}(M)$ before being recycled as many times as fixed by the variable 'cycl'.

Impairments may be introduced by means of the vector **imp**, which consists of [sigmR sigmComp sigmC offOpA]. The flash converter discrimination levels are controlled by means of a resistive divider and a bank of comparators. 'sigmR' defines the standard deviation of the divider resistances. The standard deviation of the comparator offsets are set by 'sigmComp'. The MDAC operates like a scaled D to A capactive converter. Therefore its errors are determined by means of the variable 'sigmC' (C stands for unit-capacitors). Since the Op Amp gain $2^{\wedge}(M)$ is controlled by the ratio representing the sum of all the MDAC capacitances over a single unit-capacitor, the interstage gain error is an implicit result of sigmC. This is why no interstage gain control is provided. The Op Amp offset is controlled by 'offOpA'.

The instructions **reflash** (N,sigmR,sigmComp,ext) and **remdac**(N,sigmC,ext) define respectively the comparison levels of the flash converter and the transfer function of the MDAC.

Since new sets of impairments are calculated every time these instructions are invoked, one should take care to call them before the cycling algorithm is initiated to keep the converter unchanged during the computations.

Transition position errors are corrected if the flash converter dynamic range is extended beyond Vref. This can be achieved by means of the variable 'ext', equal to 1 when the dynamic range of the flash is bound to ± 1, and 2 when its limits are extended to ± 2.

Two types of recycling converters are available: the non-restoring one and the RSD. The instructions **recycl**, **reflash** and **remdac** are duplicated for the RSD mode simply by adding the characters **RSD** to the above instructions.

4) Testing tools

A number of tools are available for evaluation of the impact of defects on the performances of D to A and A to D converters. These are controlled by instructions enabling them to perform display functions, spectral analysis and code density tests.

Visualization

The usual MATLAB instructions suffice to display data. The instruction **visut**(q) shortens writing programs for it infers the horizontal scale automatically from the length of the data 'q' and plots as many curves as there are columns in the matrix 'q'.

Static characteristics

The instructions below are used to display static characteristics.

The instruction **datrsf**(q,N,S) displays the transfer characteristic(s) and INL–DNL curve(s) of N-bit D to A converters whose output data are described by the column vector or matrix 'q'. The INL and DNL characteristics are plotted after the pause. The vertical scales are set automatically (green lines +/− 1/2 LSB, red lines +/− 1 LSB). 'S' defines the log of segments number when considered. When not specified, S is equal to zero.

The instruction **adtrsf**(in,q) displays the static transfer characteristic of A to D converters (except Delta–Sigma) and INL–DNL plots after the pause. The column vector 'in' is the analog ramp applied to the converter and 'q' the corresponding output. Midpoints are shown as long as the resolution is low to avoid the risk of cluttering, the linear regression derived from these points is always displayed however. The most extreme points of the transfer characteristic are never taken into account for linear regression computations. Neither saturation nor the overall offset of the transfer characteristic influence the 'gain' of the converter in this manner. The magnitude of the gain derived from the linear regression is printed in the command window in the same time as the overall offset of the transfer characteristic. The same kind of flexibility is not offered for D to A converters because their transfer characteristics are always normalized before characteristics are plotted.

Dynamic characteristics

Two kinds of dynamic tests are currently available: spectral analysis and code density test.

The instruction **avrsptr**(q) computes the spectrum averaged over the 'St' columns of the data contained in the matrix 'q'. Making 'St' equal to 5 or 10 reduces the amount of random noise superimposed on the individual spectra. It is still possible to look at the individual spectra nevertheless by making use of the instruction **sptr**(q).

Spectral tests are relevant for both D to A and A to D converters, whereas code density tests are restricted to A to D converters exclusively. The instruction **coden**(Ampl,q) displays the code density test and compares the result to the ideal probability density of the input sine wave.

5) Examples

The list below illustrates a number of examples that can be performed with the accompanying toolbox. All the examples are invoked via instructions **explxx.m** where xx stands for one of the file numbers below.

expl01.m Static characteristics of scaled and segment D to A converters

This file displays the static transfer characteristic as well as INL and DNL curves of scaled and segment converters. The INL–DNL characteristics are needed when the number of bits N is larger than 8, because the fine granularity of the static transfer characteristic does not allow for the resolution of individual steps anymore. It is recommended that you start with a scaled converter making 'S' and 'sigmS' equal to zero before considering segment converters. Making use of 'St' offers the possibility to compare the characteristics of several converters having distinct impairments but the same sigm's.

The performances of R2R and unit-elements D to A converters can easily be compared. Those of R2R converters are worse that those of unit-element converters with similar *sigm*'s. The R2R sigma must match the class of the converter whereas unit-elements tolerate sigmas one order of magnitude larger. Hence, the choice is either small numbers of precise resistors or large numbers of less precise unit-elements.

For segment converters, mismatches affecting the references that control the segments are easily spotted. The INL tends to look like a broken line, which reflects the impact of 'sigmS'. The DNL is always a lot better for it reflects the DNL of the internal lower-resolution converter only. To improve the INL, the tolerances of the current sources must be drastically narrowed. Mismatches should be at least one order of magnitude smaller than those of the internal converter. Notice that when a larger number of segments with a lower resolution are considered, but the resolution is kept the same, the INL improves slightly. The chances that segment references balance each other to a larger extend are enhanced indeed (assuming of course that errors obey a zero median Gaussian distribution).

Unit-elements thermometer coded D to A converters may be tested also. As expected, their DNL is always excellent whichever INL.

exmpl02.m Histograms and cumulative curves of INL and DNL

This file is intended to give a more precise picture of the performances of scaled and segment converters. A substantially larger number of converters is considered (St may be equal to 100 or even larger). The max, INL and DNL errors are recorded for every converter. Their histograms are displayed versus the tolerances expressed in LSBs plotted horizontally and, after the pause, the cumulative curves.

The interaction between the INL and the DNL depends upon the type of converter considered. In segment converters the DNL is decoupled from the INL. Thermometer coded converters always exhibit excellent DNL performances.

exmpl03.m Spectral analysis of scaled and segment D to A converters

This file is about the dynamic performances of scaled and segment D to A converters. Spectra offer a different but very efficient way to assess the impact of imperfections on the performances of converters. An increase of the background noise with respect to quantization noise means that the DNL may be the limiting factor as far as accuracy, whereas harmonics point towards the INL. It is possible to evaluate intermodulation products by inputting two sine waves with distinct but close frequencies instead of one.

exmpl04.m Static characteristics, INL and DNL of A to D converters

The transfer characteristic and INL–DNL curves (obtained after the pause) of A to D converters are displayed in this file (except for Sigma–Delta converters). Only one characteristic is plotted at a time because the computation times that are needed are longer than those of D to A converters.

Algorithmic converters

The variable 'cycl' that defines the number of times the algorithm is supposed to run represents also the number of bits when the converter is ideal. The impact of gain mismatch and offsets can be analyzed. In RSD converters, when Vmin and Vmax are changed from zero to non-zero values, the single comparison level step splitting of converters running the same number of cycles is clearly visible.

Flash converters

Defects result from mismatches of the resistors defining the reference divider and offsets of the comparators. Although fast by essence, flash converters are slow once they are simulated by means of computers for their behavior is analyzed step by step. The computation time is unnoticeable below 10 bits however.

Recycling converters

Transition position errors and transition magnitude errors may be introduced by means of the variables sigmR, sigmComp and sigmC. The transition errors are compensated when the flash converter dynamic range is extended. When 'ext' is equal to 2 the dynamic range widens till ± 2 whereas when 'ext' is equal to 1, the dynamic range resumes to ± 1.

The display shown in exmpl05.m illustrates the way errors affect the behavior of the flash converter and the MDAC. It is advised to run this file before the others to understand the mechanisms controlling accuracy, especially the auto-correction algorithm.

The impact of errors affecting the MDAC depends greatly on the number of bits M resolved per cycle. Errors are divided indeed by the interstage gain $2^{\wedge}(M)$. One can check this feature by comparing the performances of two converters with different numbers of cycles and bits but with the same resolution.

exmpl05.m Two-step A to D cyclic converter waveforms

This file displays the output of the flash converter after each cycle as well as the reconstructed output code words. Errors affecting the reference divider introduce transition position errors that are easily spotted since the transitions do not coincide with the equally spaced vertical grid. The amplified differences between the input data and output of the flash converter encompass a scale wider than resolvable by the flash. Once the scale is expanded (ext = 2), saturation is avoided however.

exmpl06.m Spectral signature of A to D converters

This file, which is the A to D counterpart of exmpl03.m, considers algorithmic, flash and recycling converters (for Delta–Sigma noise shapers, the reader should refer to exmpl09.m).

To perform an intermodulation test, use two sine waves instead of one and remember that the converter input dynamic range should not exceed the range between −1 and +1 to avoid the saturation of the quantizer.

The instruction **harm** lists items such as the magnitude of the fundamental, the quantization noise floor density, the SFDR and the magnitudes of the main harmonics.

exmpl07.m Code density test of A to D converters

The code density test can be performed considering the same converters as those listed in exmpl04.m. The input is a pure analog sine wave whose probability density curve is plotted in the same graph as the code density test. The ratio of the two curves is displayed below.

Artefacts affect the code density test when the number of samples and bins fixed by the resolution are commensurable. The variable P defining the number of samples should always exceed the number of bits by at least 4.

exmpl08.m SNR versus magnitude of A to D converters

This file applies to the same converters as those listed in example 04 (Sigma–Delta converters are treated separately). The plot, which represents the Signal-to-Noise ratio of the output data versus the magnitude of the input signal, is currently used to evaluate the Effective Number of Bits (ENOB) as well as the dynamic range. The input signal is the usual **sinput** instruction, every column being pre-multiplied by a distinct coefficient accordingly to a log-scaled law.

It is interesting to compare the SNR plots of single- and two-level algorithmic converters. Owing to their additional bit, the SNR curve of the ideal RSD converter lies always 6 dB above that of the ideal single-level converter running the same number of cycles.

In the single-level converter, gain mismatch shifts the SNR curve downwards, parallel to itself, whereas in the RSD converter the SNR is not affected at low level but bends gradually while the input increases. The RSD converter is therefore more or less similar to a compander.

exmpl09.m Spectral content of noise shapers

This test brings to the fore some of the main features of noise shapers, namely the shape of noise density spectra versus the order of the noise shaper and the number of quantization bits. The three first noise shapers are those depicted in Fig. 7.13. The last is the fourth order noise shaper described in the paper of Chao, K. C.H., Nadeem, S., Lee, W.L., Sodini, Ch., in the *IEEE Trans CAS* **37**, (3), March 1990, 309–18.

It is possible to introduce impairments like Op Amp leakage and saturation as well as errors affecting the A to D and D to A internal converters in the three first noise shapers. When the gain G of the accumulators is low (e.g. 0.95, which is equivalent to an Op Amp gain of 20) the low frequency noise floor increases. Mismatches in the D to A converter (sigmDA = 0.001) introduce low frequency noise and harmonic distortion that is clearly visible when the order of the noise shaper is large (typ. 3) and the number of samples exceeds $2^{\wedge}(12)$. Very larger sigmADs are needed to produce similar performance degradation for the A to D converter does not belong to the feedback branch. When single bit quantizers are considered, sigmAD and sigmDA do not affect the performances at all.

exmpl010.m Magnitude of signals after integrators and quantizer

The way impairments affect the three dtsgX converters of exmpl09.m is easily understood if time domain plots representing the signals delivered by the integrators and quantizer are contemplated. It is recommended to start this example with a multi-bit noise shaper (N being equal to 3 or 4) for easing the interpretations.

Noise shapers tend to increase the high frequency quantization noise and lower the amount of low frequency noise. For instance in the 3d order noise shaper, more noise is being superimposed on the sine waves as we move from the first to the third integrator.

With small numbers of quantization bits, e.g. one, the amount of quantization noise is large. Consequently the integrators tend to output large signals that may lead to saturation. A single-bit noise shaper whose input sine wave magnitude is close to 0.5 saturates almost immediately. Long sequences of data stuck at +1 or −1 are experienced as a result. Owing to the large signals before the quantizer, a lot of time can be needed before re-entering the dynamic range of the latter. This introduces low frequency components that produce a large quantity of harmonics. It is possible to shorten somehow the duration of the saturated periods by keeping the dynamic range of the signal applied to the quantizer within more narrow limits. Another way to enlarge the dynamic range is to extend the quantizer dynamic range slightly beyond plus and minus Vref.

exmpl011.m SNR versus magnitude of noise shapers

Same test as in exmpl08.m but the noise power density is integrated over the baseband Nc instead of the half-sampling frequency. This is a simple way to simulate the operation of an ideal boxcar decimator. The optional variable Nc fixes the so-called cut-off frequency. Beware of the fact that if Nc is too small, few spectral contributions enter into the spectral power count and the erratic character of the SNR plot increases. A large Nc improves the SNR plot but requires more computation time to keep the OSR unchanged.

When Nc is not specified the baseband is automatically equal to the sampling frequency divided by 2.

exmpl012.m Second order accumulate-and-dump decimator

The spectral signature after a sinc^2 filter is shown in this example. The file makes use of the data stored in the QUANT file after running the noise shaper considered in exmpl09.m. There is no need thus to repeat exmpl09.m when re-running this example to consider the impact of the downsampling rate of the decimator.

The decimated output signal is shown after the pause.

exmpl013.m Robertson plot

This file shows the Robertson plot of single- or multi-level cyclic (algorithmic) converters. The input is a scaler. The test is intended to illustrate how errors affect the performances.

List of converter toolbox functions

- adtrsf

    Purpose
    static transfer characteristic, gain, offset, INL and DNL of A to D converters

    Synopsis
    adtrsf(in,q)

    - in   'analog' input ramp (single column vector)
    - q   quantized output data (except Delta–Sigma)

    Description
    adtrsf(in,q) displays the static transfer characteristic of A to D converters and INL and DNL after the pause. Midpoints are shown as long as the numb. of bits is smaller than 6. Gain and offset are printed in the first plot. The vertical scale of the INL and DNL displays is comprised between $+$ and -2 LSB. The green lines define tolerances of $+/-$ 1/2 LSB, the red dashed lines $+/-$ one LSB.

- algor1

    Purpose
    Single level cyclic A to D conversion function

    Synopsis
    algor1 (in,cycl,offOpA,offComp,gain)

    - in   'analog' input column vector or matrix
    - cycle   number of cycles (when ideal = numb of bits)
    - offOpA   Op Amp offset
    - offComp   Comparator offset
    - gain   error factor of the gain (ideally 1)

    Description
    algor1(in,N,A,offOpA,offComp,gain) computes the analog counterparts of the output words delivered by single level algorithmic converters. The output data are ordered in the same way as the input data. Both I/O dynamic ranges are normalized from -1 to $+1$.

- algor2

    Purpose
    two-level (RSD) cyclic A to D conversion function

List of converter toolbox functions | 243

Synopsis
algor2(in,cycl,Vmin,Vmax,offOpA,gain)
algor2(in,cycl,Vmin,Vmax,offOpA,gain,corr)

- in   analog input column vector or matrix
- cycl   number of cycles (when ideal = numb. of bits + 1)
- V_{min} lower comparison level (ideal − 0.25)
- V_{max} upper comparison level (ideal + 0.25)
- offOpA   OpAmp offset
- gain   error factor of the gain (ideally 1)
- corr   transfer characteristic offset compensation (optional)

Description
algor2(in,cycl,Vmin,Vmax,offOpA,gain,corr) computes the analog counterpart of the output words delivered by two level (RSD) algorithmic converters. The comparison levels are defined by Vmin and Vmax. The output data are ordered in the same way as the input data. The I/O dynamic ranges are normalized from −1 to +1. The optional variable 'corr' controls the Op. Amp. offset compensation algorithm. When not specified, no offset correction takes place. When specified (any variable) the sign of the Op.Amp. input is reversed after the first cycle to achieve the offset compensation.

- **avrsptr**

Purpose
spectrum of the 'q' matrix data averaged over St columns (see also the instruction **sptr**)

Synopsis
avrsptr(q)

- q   input data ('analog' or 'digital')

Description
avrsptr(q) displays the fft averaged over the St columns of the matrix 'q' of the hanning windowed data q. To ease fft calculations, the length of the input data (number of rows) is rounded automatically to the nearest power of two by dropping the first samples. The objective is to eliminate the samples introduced by means of the variable D in order to dissolve transients affecting Delta Sigma noise shapers.

- **binm**

Purpose
binary conversion matrix used to compute look-up tables of scaled and segment D to A converters

Synopsis
binm(N)

- N   number of bits

Description
binm(N) creates a binary matrix whose rows represent all possible N bit code words from [0 0 . 0] to [1 1 ... 1]. The matrix product **binm**(N)***binw**(Typ,N.sigm) is a column vector which represents the look-up table of any D to A converter whose binary scale is specified by 'Typ', 'N' and 'sigm' (see also **binw**(Typ,N,sigm)).

Example
a = **binm**(3)

0	0	0
0	0	1
0	1	0
0	1	1
1	0	0
1	0	1
1	1	0
1	1	1

- **binw**

Purpose
column vector representing the ordered weights of any R2R or unit-elements binary scaler specified by Typ, N and sigm. (see also **binm** and **trm**)

Synopsis
binw(Typ,N,sigm)

- Typ   '1' for R2R; '2' for unit-elements converters
- N   number of bits
- sigm   standard deviation of resistances (Typ = 1) or unit-elements (Typ = 2)

Description
binw(Typ,N,sigm) computes a column vector that represents the weights of any R2R or unit-element binary scaler specified by Typ, N and sigm. Typ '1' referes to R2R converters, '2' to unit-elements converters. N is the number of bits. Sigm is the standard deviation of every resistance or unit-element. A zero median Gaussian distribution is assumed for all parts. Weights range from 0.5 till 2^{-N}.

Example
b = **binw**(2,3,0)

    0.5000
    0.2500
    0.1250

The operation **binm**(3)*****binw**(2,3,0) yields the 'analog' counterpart of the binary output of an ideal 3 bit unit-elements D to A converter whose range extends from 0 to 1.

0.0000
0.1250
0.2500
0.3750
0.5000
0.6250
0.7500
0.8750

- **coden**

  Purpose
  code density test

  Synopsis
  coden(Ampl,q)

  - Ampl   magnitude of the 'analog' input sine wave applied to the A to D converter under consideration
  - q   quantized output data from the A to D converter

  Description
  coden performs the code density test of the quantized data q. The graph shown above displays the result of the test, and compares it to the probability density of the ideal analog input sine wave (this is why the variable 'Ampl' is needed). The plot below shows the ratio of the code density test over the ideal probability density curve.

- **dal**

  Purpose
  computes the analog output data of **binary** scaled (R2R and unit-elements) D to A converters and segment converters

  Synopsis
  dal(in,N,Typ,sigm)
  dal(in,N,Typ,sigm,S,sigmS)

  - in   digital input (single column vector or matrix)
  - N   number of bits
  - Typ   '1' for R2R, '2' for unit-element **binary** scalers
  - sigm   stand dev of every element that belongs to the scaler
  - S $\log_2$ of the number of segments (optional)
  - sigmS   stand dev of the segment reference sources (optional)

List of converter toolbox functions

Description

dal(in,N,Typ,sigm,S,sigmS) computes the analog output of N-bit **binary** scaled and segment converters. The digital input is the column vector or matrix 'in'. Since 'in' must be the 'analog' counterpart of the 'digital' data inputted to the converter, the intruction **qtz** should be used to construct 'in'. The optional variables 'S' and 'sigmS' define respectively the number of segments 2^S and the stand deviation of every segment reference. The default values of 'S' and 'sigmS' are zero, turning the converter into a scaled converter when not defined.

- **da2**

Purpose

computes the analog output data of **thermometer coded** unit-elements D to A converters

Synopsis

da2(in,N,sigm)

- in   digital input (single column vector or matrix)
- N   number of bits
- sigm   stand dev of every unit-element

Description

da2(in,N,sigm) computes the analog output of N-bit **thermometer scaled** converters. The digital input is the column vector or matrix 'in'. Since 'in' must be the 'analog' counterpart of the 'digital' data inputted to the converter, the intruction **qtz** should be used to construct 'in'.

- **datrsf**

Purpose

static transfer characteristic, INL and DNL of D to A converters

Synopsis

datrsf(q,N)
datrsf(q,N,S)

- q   output data from the D to A converter
- N   number of bits
- S $\log_2$ of the number of segments (*optional*)

Description

datrsf displays the transfer characteristic(s) and INL DNL curves of N-bit, scaled and segment D to A converters whose output is given by the column vector or matrix 'q'. The INL and DNL plots are displayed after the pause. The vertical scales are set automatically (green lines +/− 1/2 LSB, red lines +/− one LSB).

List of converter toolbox functions | 247

- **dcim**

  Purpose
  sinc^2 decimation filter

  Synopsis
  dcim(q,b)

  - q   data from the noise shaper
  - b   downsampling rate

  Description
  dcim performs the sinc^2 decimation of the data contained in the column or matrix 'q' representing the output of a noise shaper. The factor 'b' defines the downsampling rate.

- **dtsg1**

  Purpose
  first order noise shaper

  Synopsis
  dtsg1(in,N,G,B,sigmAD,sigmDA)

  - in   analog input column vector or matrix
  - N   number of quantization bits
  - G   gain of the accumulators
  - B   accumulator symmetrical saturation bounds
  - sigmAD   stand dev of the flash A to D converter resistors
  - sigmDA   stand dev of the unit-elements of the D to A scaled conv.

  Description
  dtsg1 implements the first order noise shaper shown in Fig. 7.13. The input may be a column vector or a matrix. The variable N fixes the number of bits of the quantizer. SigmAD and sigmDA are the stand dev that control the imperfections of the A to D (flash) and D to A (unit-elements) converters. The variable G sets the gain of the accumulators (nominally 1) while B resolves the upper and lower saturation bounds.

- **dtsg2**

  Purpose
  second order noise shaper

  Synopsis
  dtsg2(in,N,G,B,sigmAD,sigmDA)

  - in   analog input column vector or matrix
  - N   number of quantization bits
  - G   gain of all accumulators feedback loop
  - B   accumulator symmetrical saturation bounds
  - sigmAD   stand dev of the flash A to D converter resistors
  - sigmDA   stand dev of unit-elements of the D to A converter

Description

dtsg2 implements the second order noise shaper shown in Fig. 7.13. The input may be a column vector or a matrix. The variable N fixes the number of bits of the quantizer. SigmAD and sigmDA are the stand dev that control the imperfections of the A to D (flash) and D to A (unit-elements) converters. The variable G sets the gain of the accumulators (nominally 1) while B resolves the upper and lower saturation bounds.

- **dtsg3**

Purpose
third order noise shaper

Synopsis
dtsg3(in,N,G,B,sigmAD,sigmDA)

- in   analog input column vector or matrix
- N   number of quantization bits
- G   gain of all accumulators feedback loop
- B   accumulator symmetrical saturation bounds
- sigmAD   stand dev of the flash A to D converter resistors
- sigmDA   stand dev of unit-elements of the D to A converter

Description

dtsg3 implements the third order noise shaper shown in Fig. 7.13. The input may be a column vector or a matrix. The variable N fixes the number of bits of the quantizer. SigmAD and sigmDA are the stand dev that control the imperfections of the A to D (flash) and D to A (unit-elements) converters. The variable G sets the gain of the accumulators (nominally 1) while B resolves the upper and lower saturation bounds.

- **flash**

Purpose
flash A to D converter (see also **flash2** and **flash2RSD**)

Synopsis
flash(in,N,sigmR,sigmComp)

- in   analog input (column vector or matrix)
- N   number of bits
- sigmR   stand dev of reference divider resistances
- sigmComp   stand dev of comparators offsets

Description

flash(in,N,sigmR,sigmComp) outputs the coded words of N-bit flash converters. The input 'in' may be a column vector or a matrix. The reference scale consists of resistors whose stand dev is given by sigmR. The comparators offsets are set by sigmComp.

List of converter toolbox functions | 249

- **flash2**

    Purpose
    flash A to D converter (see also **flash** and **flash2RSD**)

    Synopsis
    flash2(in,r,ext)

    - in   analog input (column vector or matrix)
    - r   reference levels (defined by the instruction **reflash**)
    - ext   extention of the dynamic range of the converter

    flash2(in,r,ext) outputs the coded words of N-bit flash converters. The input 'in' may be a column vector or a matrix. Unlike in the instruction **flash**, the resistive divider is resolved by means of a separate instruction **reflash**(N,sigmR,sigmComp,ext), to allow the repeated use of the same divider without changing its impairments. This is needed when multibit recycling converters are simulated. The variable 'ext' (respectively equal to 1 or 2) extends the dynamic range of the converter from $\pm$ 1 till $\pm$ 2. Extending the conversion scale compensates the effects of divider resistors mismatches and comparators offsets which otherwise affect the resolution of multibit recycling converters.

- **flash2RSD**

    Purpose
    RSD flash A to D converter (same as **flash2** but for RSD converters)

    Synopsis
    flash2RSD(in,r)

    - in   analog input (column vector or matrix)
    - r   reference levels (defined by the instruction **reflashRSD**)

    flash2RSD(in,r) outputs the coded words of N-bit RSD flash converters. The input 'in' may be a column vector or a matrix. Unlike in the instruction **flash**, the resistive divider is resolved by means of a separate instruction **reflashRSD**(N,sigmR,sigmComp), which allows the repeated use of the same divider without changing its impairments. To be used with multibit RSD recycling converters.

- **harm**

    Purpose
    lists the noise floor power density, fundamental and harmonics in descending order as well as the SFDR of the spectrum defined by S

Synopsis
harm(S)
harm(S,b)

- S   spectral distribution (e.g. data from **avrsptr**)
- b   (*optional*) limits the harmonic content to those rays only which exceed the noise floor defined by b (dB). The default value of b is 20 dB.

Description
harm(S,b) finds the fundamental, measures its magnitude and evaluates the noise floor in the spectral region comprised between the fundamental and the second harmonic. Once the largest harmonic is spotted, its ampltude and rank are evaluated. The magnitude difference with respect to the fundamental yields the SFDR. The algorithm displays also in descending order the harmonics exceeding the noise floor by b dB.

Example

rank	magn (dB)	magn referred to fundam (dB)	
—	−113.6190	—	← noise floor
1.0000	−9.0296	0	← fundamental
3.0000	−84.2872	−75.2576	← SFDR
2.0000	−86.4661	−77.4365	← other harm
4.0000	−93.1054	−84.0758	
6.0000	−95.1539	−86.1243	

- **qtz**

Purpose
ideal A to D converter (quantizer)

Synopsis
qtz(in,N)

- in   analog input column vector or matrix
- N   number of bits

Description
qtz(in,N) outputs the N bit quantized counterpart of the continuous input 'in', which is a column vector or a matrix. The input/output dynamic range is comprised between −1 and +1. The instruction implements the cascade combination of an ideal A to D converter followed by an ideal D to A converter.

- **recycl**

Purpose
recycling multibit A to D converter (see also **recyclRSD**)

Synopsis
recycl(in,M,cycl,imp,ext);

List of converter toolbox functions | 251

- in   analog input column vector or matrix
- M   number of bits of the flash and MDAC converters.
- cycl   number of cycles
- imp   impairments vector = [sigmR sigmComp sigmC offOpA]
- ext   extension of the reference scale

Description

recycl(in,M,cycl,imp,ext) computes the output of recycling M-bit A to D converters. The resolution is equal to M*cycl. Impairments are set the row vector **imp**. The variable 'ext' (respectively 1 or 2) extends the dynamic range of the converter from ± 1 till ± 2. When 'ext' is equal to 2, the impairments from the resistive divider and comparators offsets are compensated.

- **recyclRSD**

Purpose

recycling multibit RSD A to D converter (see also **recycl**)

Synopsis

recyclRSD(in,M,cycl,imp);

- in   analog input column vector or matrix
- M   number of bits of flash and MDAC converters.
- cycl   number of cycles
- imp   impairments vector = [sigmR sigmComp sigmC offOpA]

Description

recyclRSD(in,M,cycl,imp) computes the output of recycling M-bit A to D converters. The resolution is equal to M*cycl+ 1. Impairments are set by the row vector **imp**.

- **reflash**

Purpose

reference divider for flash converter (see also **reflashRSD**)

Synopsis

reflash(N,sigmR,sigmComp,ext);

- N   number of bits
- sigmR   stand dev of the divider resistances
- sigmComp   stand dev ot the comparators offsets
- ext   extension of the reference scale

Description

reflash(N,sigmR,sigmComp,ext) sets the reference scale of flash converters whose resistance mismatches and comparators offsets are controlled respectively by the variables 'sigmR' and 'sigmComp'. The instruction is intended for recycling multibit converters where the flash divider must remain unchanged while cycling is taking place. The variable 'ext' (respectively 1 or 2) extends the dynamic range of the converter from ± 1 till ± 2. Making 'ext' equal to 2 compensates the impairments from the flash references which otherwise affect the accuracy of the converter.

- **reflashRSD**

Purpose
reference divider of flash RSD converter (see also **reflash**)

Synopsis
reflashRSD(N,sigmR,sigmComp);

- N   number of bits
- sigmR   stand dev of the divider resistances
- sigmComp   stand dev ot thr comparators offsets

Description
reflashRSD(N,sigmR,sigmComp) computes the reference scale of RSD flash converters whose resistance mismatches and comparators offsets are controlled respectively by the variables 'sigmR' and 'sigmComp'. The instruction is intended for recycling multibit converters where the flash divider must remain unchanged while cycling is taking place.

- **remdac**

Purpose
MDAC for recycling A to D converters (see also **remdacRSD**)

Synopsis
remdac(N,sigmC,ext);

- N   number of bits
- sigmC   stand dev of the MDAC unit-elements.
- ext   extension of the reference scale

Description
remdac(N,sigmC,ext) computes the unit-elements of the MDAC used in multibit recycling converters. The parameter 'sigmC' controls the capacitors standard deviation and the interstage gain, which is listed in the command window. The variable 'ext' (respectively 1 or 2) extends the dynamic range of the converter from ± 1 till ± 2.

- **remdacRSD**

Purpose
MDAC for recycling A to D RSD converters (see also **remdac**)

Synopsis
remdacRSD(N.sigmC);

- N   number of bits
- sigmC   stand dev of the MDAC unit-elements

List of converter toolbox functions | 253

Description
remdacRSD(N,sigmC) computes the elements of the MDAC to be used in multibit RSD recycling converters. The parameter 'sigmC' controls the capacitors standard deviation and defines the interstage gain, which is listed in the command window.

- **sat**

Purpose
magnitude limiter

Synopsis
sat(in,B)

- in   input signal (column vector or matrix)
- B   saturation boundary

Description
sat(in, B) duplicates 'in' when the signal is comprised between $-B$ to $+B$. Outside the output is stuck at $-B$ or $+B$.

- **sinput**

Purpose
creates a matrix of 'analog' input signals

Synopsis
sinput(St,Np,P)
sinput(St,Np,P,D)

- St   number of sine wave columns
- Np   number of entire periods in each column.
- P   the number Ns of 'analog' samples $= 2^{\wedge}(P)$
- D   adds D samples to Ns (optional)

Description
sinput(St,N,P,D) generates a matrix whose columns represent sine waves with the same number of periods but random phases. The instruction is intended for the dynamic testing of A to D converters. The number of sine waves is controlled by the variable St. When larger than 1, St lessens the computation noise of the fft when the **avrsptr** instruction is used (this is why random phases are contemplated). The number of rows Ns is equal to $2^{\wedge}(P)$. When state variables are contemplated like in noise shapers, the optional variable D is used to add D more samples to Ns to ignore transients. The D first samples of the output are erased automatically when computing the fft. The default value of D is zero. Cascading the **sinput** and **qtz** instructions yields the same data for D to A converters.

- **snrplot**

Purpose
signal-to-noise plot of A to D converters

Synopsis
snrplot(q,Ampl,Np)
snrplot(q,Ampl,Np,Nc)

• q	matrix representing the output data from an A to D converter.
• Ampl	row vector controlling the magnitudes of the analog input sine waves applied to the D.U.T. (horiz. scale of SNR plot)
• Np	number of entire periods in every input column
• Nc	optional variable restricting the baseband to Nc instead of Ns/2 (for delta-sigma converters)

Description
snrplot(q,Ampl,Np,Nc) displays the signal-to-noise plot of A to D converters versus the magnitude of the input. The Device Under Test (D.U.T.) must be fed by a **sinput** matrix whose St columns are pre-multiplied by the factors stored in a St-long logspaced 'Ampl' row vector. The noise power is integrated from (almost) DC till either the half sampling frequency Ns/2 (default value of Nc) or Nc when specified. Nc allows to reduce the noise power bandwidth in Delta–Sigma converters. The choice of Nc is equivalent to the introduction of an ideal low-pass boxcar decimator whose cut-off frequency is controlled by Nc.

• **sod**

Purpose
4th order noise shaper

Synopsis
sod(in)

• in	input column vector or matrix

Description
sod(in) displays the spectral noise density of the fourth order Delta-Sigma noise shaper described by Chao, K.C.H., Nadeem, S., Lee, W.L., Sodini, Ch., in *IEEE Trans CAS*, 37, March 1990, 309–318.

• **sptr**

Purpose
Individual spectra of the data contained in the St columns of matrix 'q' (see also **avrsptr**).

Synopsis
sptr(q)

• q	input data ('analog' or 'digital')

List of converter toolbox functions | 255

Description
sptr(q) displays the individual fft's of the St columns of the matrix 'q' (hanning window).

- **trm**

Purpose
Conversion matrix used to compute look-up tables of scaled D to A converters whose binary weights are described by binw. See also **binw**(Typ,N,sigm).

Synopsis
trm(N);

- N   number of bits

Description
trm(N) creates a matrix whose rows represent all possible N bit RSD code words in ascending order from -1 till $+1$. The matrix product **trm**(N)*****binw**(typ,N,sigm) creates a column vector which is the two-quadrant look-up table of D to A converters whose binary scale is specified by 'Typ','N' and 'sigm'.

Example
a = trm(3)

-1	-1	-1
-1	0	-1
0	-1	-1
0	0	-1
0	0	1
0	1	1
1	0	1
1	1	1

b = binw(2,3,0)

0.5000
0.2500
0.1250

a*b

−0.8750
−0.6250
−0.3750
−0.1250
0.1250
0.3750
0.6250
0.8750

List of converter toolbox functions

- **tst**

  Purpose
  finds the resolution and 'type' of quantized data.

  Synopsis
  tst(q)

  - q   quantized input data

  Description
  tst(q) finds the number of bits 'N' and the type 'm' of the quantized data 'q'. The instruction is used when the resolution and the kind of A to D converter are unknown. The type 'm' information describes the group to which the data 'q' belong. Type 1 characterizes quantizers without a zero step like in **algor1**, while type 2 designates those with a zero step like in **algor2**.

- **visut**

  Purpose
  visualization of data versus time

  Synopsis
  visut(in)

  - in   input data (column vector or matrix)

  Description
  visut(in) displays the individual columns of 'in' versus time.

Index

accumulate-and-dump filter 175
acquisition time, *see* aperture time
algorithmic converter 94–6
   voltage mode 99
   current mode 100
all-MOS D to A converter 34–6
anti-alias filter 80, 161–2
aperture time 77
auto-calibration, auto-correction 36–8, 87–9, 170

band-pass Delta-Sigma converter 172–4
bandwidth, converter 23
binary data selection 28–32
binary, fractional expansion
   most significant bit (MSB) 1
   least significant bit (MSB) 1
bubble, thermometric code 190

C.D.T. code density test 19–22
capacitive D to A converters 44–7
   layout 47–9
capacitive divider 44
capacitor, integrated 41–3
cascaded Delta-Sigma converters, *see* MASH
CCITT specifications 92, 124
charge injection, sharing 72–5, 209
charge redistribution converter 83–5
code; *see also* test
   binary 1
   code correction 198, 206
   missing code 10
   target code 10
   thermometer code 39–41
codec, compander 65, 90–3, 121, 124

comparator; *see also* offset 85–7
   flash converter 186–9
copier, current 67–72
correction strategies 198–200
cumulative histograms 29, 40, 42, 61
cyclic converters, *see* algorithmic converters

D.N.L. differential-non-linearity 6
D.U.T. device under test 10
D.W.A. data weighted averaging 170
Decimator, decimation 175–7
Delta-Sigma A TO D conv; *see also* noise shaper 139–77
   baseband 139
   generic 146–7
   linear approximation 142–6
   principle 139–42
Delta-Sigma D to A converters 178–82
distortion, harmonic distortion 14–7
doublet 32
down-sampling 147, 175–7
droop, voltage 70
dual-slope converters 127–30
dynamic current division principle 54–9
dynamic current matching 54–9

E.N.O.B. effective number of bits 18

fft fast Fourier transform 14–19
flash 183–6
   effects of offset 188
   offset averaging 188
floating point RSD converter 121–4
folder circuits 211, 213
folding converters 209–12
   interpolating folding converters 212–17

gain error 6
  effects in algorithmic converters 102–9
  correction techniques 109–13
glitch 31
G.V.O. gate voltage overdrive

Hanning window, spectral leakage 18

idle-tones 161
I.N.L. integral-non-linearity 5
intermodulation 16, 208
intermodulation products 16
interpolation 178–80
interstage gain 104–197

jitter 79, 161, 185

kickback noise 189

limit cycle 157
linearity
  analog-to-digital 5–8
  digital-to-analog 8–10
loop filter, Delta-Sigma 142, 146, 150, 161, 163

MASH converters 170–2
metastability 190–1
  digital correction 190
midpoint, transfer characteristic 9
mismatch
  capacitor 43–4
  resistors 24–26
  transistor 49–50
modulator, see noise shaper
multi-bit Delta-Sigma converters 168–70
multistep converters, see subranging, recycling and pipelined converters
multistep single-bit converters 102

noise in algorithmic converters 117
noise shaper 150–55
  noise shaping 141–2
  synthesis 164–8
non-monotonicity 7

offset
  comparator 186–9
  correction in algorithmic converters 116–7
    in charge transfer rampconverters 133
    in ramp converters 129–30
  effects in algorithmic converters 113–6
  transfer characteristic 6
O.S.R. oversampling rate 141
oversampling 139

parasitics, capacitive 43
pedestal 30, 44
pipelined converters 195–6
position dependent time skew 185
power density spectrum, quantized signal 138
probability density 19, 135

quantization noise 14, 135–9
  spectrum 138
quantizer 135

R2R D to A converters 32–38
R2R ladder network 26
ramp-function converters 127–33
  charge ramp function converters 131–3
randomization, mismatch 51, 170
read-out circuits for D to A converters
  MOS 30, 52
  bipolar 31
recycling converters 195
residue, remainder 94, 193; *see also* quantization noise
resolution 1, 17
Robertson plot 96–8, 108, 113–17, 119, 204
  signature, trajectory 98
root-locus 157

RSD algorithm
   current mode 125–6
   voltage mode 118–24
RSD multistep converters 203–9

sampling 76, 80–1
scaling
   capacitor 41–49
   resistor 24–38
   transistor 49–50
   unit-elements 38–41
segment converters 59–62
   current mode 62–3, 65–7
   voltage mode 64–6
S.F.D.R. spurious free dynamic range 15
S.H. sample-and-hold 76–80, 209
signal-dependent delay 189
signal feedthrough in S.H. 79
signal to noise-plus-total-harmonic-
     distortion 18
signal-to-noise ratio of codecs 92
sinc filter 175
skew, time 185
S.N.R. signal-to-noise ratio 16, 18, 79, 139
spectral leakage 18
spectral signature 14–6

stability of noise shapers 157–60, 168
stray capacitance, *see* parasitics
sub-ranging converters 192–3
successive approximation algorithm 81–3

test
   beat frequency test 14
   code density test, C.D.T. 19–22
   fast Fourier transform test fft 14–9
   ramp test 11
   resolution 1, 17, 154–5
   servo loop test 12
tolerance tables
   resistors 25
   capacitors 43
   transistors 49–50
transfer characteristic 1
transistor scaled D to A converters 50–3
transition point 3
transition magnitude errors 196
transition position errors 196
two-step converters 193–5

unit-element scalers 38–41